MEDIA AND COMMUNICATIONS - TECHNOLOGIES, POLICIES AND CHALLENGES

CONTEMPORARY RISK COMMUNICATION

ELEMENTS WITH APPLICATIONS IN A NANOTECHNOLOGY CONTEXT

MEDIA AND COMMUNICATIONS - TECHNOLOGIES, POLICIES AND CHALLENGES

Additional books in this series can be found on Nova's website
under the Series tab.

Additional E-books in this series can be found on Nova's website
under the E-book tab.

NANOTECHNOLOGY SCIENCE AND TECHNOLOGY

Additional books in this series can be found on Nova's website
under the Series tab.

Additional E-books in this series can be found on Nova's website
under the E-book tab.

MEDIA AND COMMUNICATIONS - TECHNOLOGIES, POLICIES AND CHALLENGES

CONTEMPORARY RISK COMMUNICATION

ELEMENTS WITH APPLICATIONS IN A NANOTECHNOLOGY CONTEXT

JON B. MADSEN
EDITOR

nova publishers
New York

For permission to use material from this book please contact us:
Telephone 631-231-7269; Fax 631-231-8175
Web Site: http://www.novapublishers.com

Library of Congress Cataloging-in-Publication Data

ISBN: 978-1-62808-464-1

Published by Nova Science Publishers, Inc. † New York

Contents

PREFACE

Chapter 1 – *Risk communication* manages how a potential hazard is represented to multiple audiences. This White Paper details the risk communication literature, its strategies, and the research evidence for what happens in a risk analysis when data is unavailable due to randomness or incompleteness. It examines these issues in the context of nanotechnology, as a case study of an emerging technology area that presents risk communication challenges. Inexpert audiences may not assess risk by using equations but rather by weighing a combination of scientific, social, and emotional factors. Understanding risk perceptions and communicating risks related to nanotechnology can pose special issues. For many people, especially people who are unaware, unfamiliar and/or uncomfortable with science, the topic of nanotechnology can be difficult to understand. In addition, for some people, nanotechnology evokes concerns related to the environment, health, safety, or other societal issues. Nanotechnology's interface with biotechnology and other sciences and the complexity of nanoscience applications, present numerous challenges to risk communicators.

Experts including scientists, engineers, communication specialists, and policymakers need a knowledge base about risk perception and a basic skill set about risk communication to better understand and enhance their ability to effectively communicate risk to the public and the media. Specifically, the public's struggle with understanding science and new technologies does not necessarily emerge from a science knowledge deficit or a lack of technological literacy but rather emanates from existing personal belief predispositions and values systems that new technologies may challenge. The following are ten points to consider when communicating risk.

Chapter 2 – Nanotechnology—a term encompassing nanoscale science, engineering, and technology—is focused on understanding, controlling, and exploiting the unique properties of matter that can emerge at scales of one to 100 nanometers. A key issue before Congress regarding nanotechnology is how best to protect human health, safety, and the environment as nanoscale materials and products are researched, developed, manufactured, used, and discarded. While the rapidly emerging field of nanotechnology is believed by many to offer significant economic and societal benefits, some research results have raised concerns about the potential adverse environmental, health, and safety (EHS) implications of nanoscale materials.

Some have described nanotechnology as a two-edged sword. On the one hand, some are concerned that nanoscale particles may enter and accumulate in vital organs, such as the lungs and brains, potentially causing harm or death to humans and animals, and that the diffusion of nanoscale particles in the environment might harm ecosystems. On the other hand, some believe that nanotechnology has the potential to deliver important EHS benefits such as reducing energy consumption, pollution, and greenhouse gas emissions; remediating environmental damage; curing, managing, or preventing diseases; and offering new safety-enhancing materials that are stronger, self-repairing, and able to adapt to provide protection.

Stakeholders generally agree that concerns about potential detrimental effects of nanoscale materials and devices—both real and perceived—must be addressed to protect and improve human health, safety, and the environment; enable accurate and efficient risk assessment, risk management, and cost-benefit trade-offs; foster innovation and public confidence; and ensure that society can enjoy the widespread economic and societal benefits that nanotechnology may offer. Congressionally-mandated reviews of the National Nanotechnology Initiative (NNI) by the National Research Council and the President's Council of Advisors on Science and Technology have concluded that additional research is required to make a rigorous risk assessment of nanoscale materials.

In: Contemporary Risk Communication
Editor: Jon B. Madsen
ISBN: 978-1-62808-464-1

Chapter 1

COMMUNICATING RISK IN THE 21ST CENTURY: THE CASE OF NANOTECHNOLOGY*

David M. Berube, Brenton Faber, Dietram A. Scheufele, Christopher L. Cummings, Grant E. Gardner, Kelly N. Martin, Michael S. Martin and Nicholas M. Temple

ACKNOWLEDGEMENTS

This work was supported by a supplement to grant #0809470 from the National Science Foundation, NSF 06-595, Nanotechnology Interdisciplinary Research Team (NIRT): Intuitive Toxicology and Public Engagement.

Logistical and additional support was received from North Carolina State University, the College of Humanities and Social Sciences (CHASS), and the Department of Communication.

We especially appreciated the support of the Communication, Rhetoric and Digital Media (CRDM) doctoral program and the Public Communication of Science and Technology Project (PCOST).

We thank the efforts of Vivian Ota Wang (National Human Genome Research Institute - National Institutes of Health of the U.S. Department of Health and Human Services (NHGRI-NIH-DHHS) and NIH Agency Representative to the National Science and Technology Council (NSTC) and the National Nanotechnology Coordination Office (NNCO), and Stephen Zehr (National Science Foundation- Science, Technology, and Society (NSF-STS) both of whom are associated with the Nanotechnology Public Engagement and Communications Work Group (NPEC) of the NSTC as Chair and Working Group Member respectively.

---

* This is an edited, reformatted and augmented version of The National Nanotechnology Coordination Office and supported by a supplement to grant #0809470 from the National Science Foundation, NSF 06-595, dated February 2010.

In addition, two colleagues read through the paper and offered their insights: A. Celeste Farr and William Kinsella, Professors in the Department of Communication, NCSU. Two CRDM doctoral students, Jason Kalin and Daniel Sutko served as copy editors as well.

All authors of this White Paper are members of PCOST.

EXECUTIVE SUMMARY

Risk communication manages how a potential hazard is represented to multiple audiences. This White Paper details the risk communication literature, its strategies, and the research evidence for what happens in a risk analysis when data is unavailable due to randomness or incompleteness. It examines these issues in the context of nanotechnology, as a case study of an emerging technology area that presents risk communication challenges. Inexpert audiences may not assess risk by using equations but rather by weighing a combination of scientific, social, and emotional factors. Understanding risk perceptions and communicating risks related to nanotechnology can pose special issues. For many people, especially people who are unaware, unfamiliar and/or uncomfortable with science, the topic of nanotechnology can be difficult to understand. In addition, for some people, nanotechnology evokes concerns related to the environment, health, safety, or other societal issues. Nanotechnology's interface with biotechnology and other sciences and the complexity of nanoscience applications, present numerous challenges to risk communicators.

Experts including scientists, engineers, communication specialists, and policymakers need a knowledge base about risk perception and a basic skill set about risk communication to better understand and enhance their ability to effectively communicate risk to the public and the media. Specifically, the public's struggle with understanding science and new technologies does not necessarily emerge from a science knowledge deficit or a lack of technological literacy but rather emanates from existing personal belief predispositions and values systems that new technologies may challenge. The following are ten points to consider when communicating risk.

First, trust must be built and sustained in all risk communication. Trust amplifies (enhances) or attenuates (dampens) perceptions of risk. It is difficult if not impossible to launch a risk communication campaign solely based on authoritative claims. Campaigns must be built on trust and "good reasons." Specifically, the public will lose trust in experts and institutions when these groups are perceived to hold and promote vested and narrow self-interests. The failure of "believe us because we are experts" campaigns may explain the increasing use of fear claims by risk communicators.

Second, risk perception differences exist between experts and inexperts. This perceptual gap is compounded by communication challenges, especially when these two groups are discussing the same topic, but not speaking or understanding the same language. Additionally, people's personal experiences and biases influence how they make sense of a world, especially when feeling powerless. Everyone uses heuristics (mental short cuts) as ways to explain how people make decisions and solve problems, especially when faced with incomplete or complex information. Although most people are neither irrational nor ignorant, these shortcuts can create biases that are not only powerful influences of perceptions and sensibilities but also resilient and irrepressible. Risk researchers have shown that these mental short cuts can attune people to automatically prefer some information over others due to previous exposures and strong emotional attachments. These short cuts can reinforce biases, which favor distressing information, confirm previously held beliefs, and discount unfamiliar data such as exposure and dosage data. Working with rather than discrediting these mental short cuts and biases will avail the risk communicator with an additional communication skill set. A communication strategy that builds on familiar sensibilities and validates rather than challenges beliefs and attitudes will be more effective.

Third, communication often occurs through intermediaries such as the media. The media prime and frame issues. During a crisis both the legacy mainstream media and digital media will focus on most negative elements of the story. More control over the message will reduce undesirable framing or passing on biased information, such as calling genetically engineered food "Frankenfood".

Fourth, while traditional media (newspapers, radio) seem to both attenuate and amplify risk perceptions, digital media, including electronic technologies (blogs, podcasts, wikis, etc.) generally tend to amplify risk perceptions. Digital media are seldom subjected to editorial oversight and generally focus on the "spectacular," even if the claim is objectively and scientifically indefensible. Mainstream media will highlight the most sensational aspects of a story, exercising editorial judgment to include different sides to the story. However blogs running stories with headlines such as "Nanotubes are like asbestos" are much less likely to have a balanced editorial slant with information and quotes from scientists as will appear in mainstream media news outlets.

Because of accessibility, digital media may be replacing traditional media as a source of information for inexpert audiences. Understanding the function of digital media and the interrelationships among communication and other social factors is essential for understanding how people perceive and make judgments about risks. Rather than transferring messages designed for newspapers and television news and posting it on the web, communication must be designed for digital media outlets, such as Facebook and blogs.

Fifth, while scientists may review risk in specific disciplinary and procedural ways, inexpert audiences use more narrative and "storied" approaches that include factors mediated by belief, experience, prejudice, authority, and social and ideological predispositions to understand risk. Non-governmental organizations have been more successful in creating and disseminating their messages because they use personalized and compelling narratives. When given heavily data-laden messages using probabilities and risk estimates, inexperts become dazed. Communicators need to use "stories" involving people and familiar situations and circumstances to reach inexpert audiences.

Sixth, advocates of new technologies must become better equipped to influence which issues become public by being proactive in public agenda building and agenda setting and shape *how* the issues are debated in public discourse. The presence or absence of an issue on the public agenda (priming) and the way that issue is presented (framing) has significant influence on how people think about potential policy options and form attitudes about technology. Both the Greenpeace's "Frankenfood" frame in Europe and the "Teach the Controversy" frame are good examples of carefully researched ways of using particular labels or phrases to present an issue that indirectly offer interpretive ways audiences should think and feel about the topic.

Seventh, the concept of risk inherently has a negative valence. Similarly, while fear can play a motivational role in public risk management, the proliferation of risk messages can produced risk fatigue (the public paying less attention to risk messages because of they are over-exposed and overwhelmed by risk messages). In this situation, people may tune out to all risk messages or focus on the "wrong" risk messages. For example, parents concerned that their child may be abducted worry more about strangers than acquaintances and immediate family members, though the latter are much more likely to commit such a crime.

Eighth, personal experiences, religious belief systems, national identities, peer groups, occupations, individual characteristics, and self-identities are powerful and influential in how risk perceptions are formed, maintained, discredited, and changed. Within these contexts, science may challenge traditional definitions of what it means to be human. For example, advanced technologies may challenge people to question their beliefs and values. This can generate an inherent disconnect between scientific progress and personal values. Because personal values systems "filter" new information, various individuals and social groups may consequently develop different and personalized understandings of the same information. It is critical for communicators to develop and utilize a systematic understanding of these processes so they can develop and foster more effective dialogue with *all* groups.

Ninth, some public engagements have targeted small, non-influential, inappropriate, and/or unrepresentative stakeholder groups. In these cases, individuals and groups targeted for public outreach, involvement, participation, and engagement activities may not be sufficiently representative to yield reliable inferential data. Purposeful and research-informed engagement strategies that interact with the public where the public already assembles, rather than in artificially built experimental settings such as consensus conferences, may be a more appropriate strategy for relevant and meaningful stakeholder group participation.

Tenth, the field of risk communication is dynamic and research-based. Unfortunately, many risk communication campaigns are reactively instead of proactively developed and "intuitive" rather than informed by social, behavioral, and communication sciences research methods and data. Communication is not simply talking eloquently. Communication involves a purposeful and developed skill set. For example, crisis communication specialists often manage communication disasters because unqualified people initially mis-communicate ineffective risk and crisis messages.

Research in risk studies and communication is ongoing. Information and its impacts do not occur in a vacuum. New theories, designs, and data are being developed to better understand media and how they affect risk perceptions and communication. For example, fuzzy set theory (Smithson and Verkuilen, 2006), a modeling tool used for assessing categorical and dimensional concepts simultaneously such as in weather and disaster prediction, may provide insight in how to better communicate about the uncertainties raised by nanotechnology.

How information is presented and framed will also influence the ways people cognitively and emotionally understand and use it. In order to determine how risk data can effectively be conveyed to different audiences, communication planning and strategies should be grounded in theory and research. Risk communicators should also work with communication professionals as often as possible to improve communication skill sets.

SECTION 1. INTRODUCTION

Definitions of risk vary by situational context and specific applications. The Italian word *riscare* means to circumnavigate cliffs; an action that can put an individual or a group at risk. The range of risks and risk perceptions can include systems risks, group risks, and individual risks with each having varying degrees of tolerance for uncertainty and vulnerability.

When referring to risk assessment, people are establishing the probability of exposure and the dose or extent of the hazard a person or ecosystem must be exposed to for negative consequences to occur. Risk management entails actions by regulators and other managers who determine what can be done to reduce exposure by adopting precautions that both a specific industry and individuals can adopt. In some instances, the risk assessment may establish that the only solution is to avoid the work environment as currently conceived. For example, occupational safety professionals may calculate risk to workers by examining hazard against the probability that workers will be vulnerable to the actual or hypothetical implications of the hazard. Here, risk is code for acute and/or chronic exposure to hazard.

Finally, those who decide how much risk is acceptable or reasonable are generally the people or entities who are responsible for exposing workers and consumers to risk rather than the workers and consumers themselves. Too often, the individuals exposed to these risks do not actively participate in deciding what level of risk (read this time as exposure) is safe and are willing to tolerate.

For some people, they defer to experts and presume they know the needs and what is best for the public. In these situations, the public asks a technical community of experts to paternalistically act in the "best" interests of the public to resolve a complex risk event. This strategy presupposes that the decision-making process is voluntarily relinquished to the experts. When a technical expert community may have a conflict of interest that confounds its objectivity, other experts including those in risk assessment and communication communities may need to more actively participate.

Ethicists should also participate by determining if, how, and when the public should be informed, especially when consent is warranted. How much exposure the public is willing to choose is a complicated determination fraught with moral and ethical questions. For example, people's personal choices may outweigh higher levels of risk exposure. When a contractor travels to war-ravaged regions despite the hazard in order to earn an exorbitant salary, he decides the compensation sufficiently outweighs the risks to his life, assuming, of course, he is competent to make that decision.

Democracy depends on public participation. In a world where complex science and technology decisions are made, a policy commitment may need to be made to protect the public from unacceptable risks. Involving members of the public in deciding what level of risk is acceptable allows participation in decision making and guards against unnecessary paternalism.

Risk assessment and risk management also refer to potential hazard exposures not limited to the workplace. For example, a variety of exposures to hazards can occur as a product is manufactured, used, and disposed. In these instances, regulators and those managing exposures may offer advice or develop guidelines for distributors, consumers, and disposal professionals to handle or remove the potential hazard from the market.

The area of risk communication addresses how potential hazard exposures are represented to general and/or targeted audiences. Generally, the field of risk communication includes: care communication, crisis communication, and consensus communication. Care communication is used when the goal is to reduce exposure to a potential hazard. Crisis communication is used when the goal is to remove susceptible populations from a hazard, often after or during exposure. Unfortunately, some crisis communication efforts have been associated with more questionable public relations practices intended to reduce the liability of an offender rather than protect the target population. Consensus communication is used when the goal is both public involvement and risk reduction. It involves a plethora of stakeholders in both the process of determining hazard and probability as well as crafting approaches, when needed, to reduce exposures.

As a scholarly area of research, risk communication uses qualitative and quantitative research methods. Qualitative approaches emphasize theory and case- based-learning to teach techniques for managing risk perception premised on hypotheses drawn from theoretical insight and the historical record. Quantitative approaches, informed by qualitative research, use experimental and empirical means to manage a risk communication process. This White Paper reflects research findings from qualitative and quantitative research data and draws from both schools but favors the latter. While recognizing that human activity is notoriously unpredictable and uncertain, we argue that risk communication can be approached in a scientific way based on sound social scientific research.

A risk algorithm, as presented in Figure 1, could be based on information derived from information about hazards to human and ecological well-being. To determine risk, information is collected about potential exposure and dosage levels to a hazard from multiple sources including regulators, industry, and the public.

A probabilistic approach to risk assessment is illustrated in the Rasmussen Report, sometimes referred to as the WASH -1400 Reactor Safety Study (1975). It studied the public risk of nuclear power plant accidents. While the algorithm presented in Figure 1 is useful in illustrating a probabilistic risk assessment, this particular approach may not be adequate for assessing, especially communicating, the risks of nanotechnology because of the complexity of establishing hazard, exposure, and dosage estimates of nanotechnology given the high levels of uncertainty associated with environmental health and safety (EHS) data sets to date. For example, there are very limited if any data available because of policy and technical challenges associated with standardizing, characterizing, and measuring nanoparticles and nanoparticle exposure. Dosage, traditionally measured by mass, is confounded by the special characteristics of nanoparticles, such as increased surface area, bioavailability, and charge. In addition, life cycle exposure data from production through disposal remain unknown. There are some similarities to other contemporary hazards. For example, exposures can be potentially acute and/or chronic. Exposure(s) may also involve mortality and morbidity to humans, other life forms, and the ecosystem. In this White Paper, we offer a risk communication focused algorithm accounting for these variables.

Risk ← Hazard (Exposure x Dose)

Figure 1. Dominant algorithm in risk.

Qualitative and quantitative research approaches to the scholarship of risk, risk perception, and communication can be divided into two main camps: a psychometric approach (Fischoff, Slovic, Lichtenstein, Reads, & Combs, 1978); and a cultural approach (Douglas & Wildavsky, 1982). (A general reading list can be found in Appendix A). The cultural approach can be further divided to identify a traditional world views approach associated with Douglas and Wildavsky, an ideological view approach with Dake (1991), an elite approach with Rothman and Lichter (1987), and a cultural cognition approach of Kahan, Slovic, Braman, & Gastil (2006).

When undertaken, research studies assessing risk perception and risk communication have not generally addressed nanoparticles and some of their unpredictable characteristics. This is slowly changing. A complete list of studies on public perception of nanotechnology can be found in Appendix B.

Risk communication is a rich field of scholarship highly influenced by psychology. The following works of Sandman, Slovic, and Sjöberg illustrate the range of risk perception and communication research.

Sandman approaches risk communication as a perceptual psychologist. He introduced a "Risk = Hazard + Outrage" formula substituting "Outrage" for "Probability" to explain the "probability neglect" bias whereby people ignore or assign either a 100% or 0% to the likelihood of future events (Sandman, 1993). **Sandman** also produced a list of the twelve components of outrage including factors such as whether a hazard is involuntary or exotic.[1]

Slovic approaches risk communication as a quantitative social psychologist. He and his colleagues have developed quantitative methods to describe risk perceptions and measured their influence on individuals, industry, and society. **Slovic's research**[2] ranges from simple and complex risk, general perception, and trust. Slovic's work is regarded in the field as a standard on public attitudes toward radioactive waste, integrated waste management, prescription drugs, cigarette smoking, automotive safety devices, blood transfusions, pesticides, biotechnology, terrorism, asteroid impacts, and nanotechnology. In addition, Slovic is a major contributor to the scholarship on extreme events, numbing, dread, stigma, fear and affect, social amplification of risk, and relationships between lay and expert estimations of risk.

Sjöberg's work is informed by social and personality psychology. He provides alternative theories to many traditionally held axioms in the field of risk communication. He uses quantitative research methods as well. Sjöberg is at the forefront of research related to trust, risk perception, and antagonism. His research has shown that antagonism and epistemic trust (such as trust in science and technology) were more important than social trust in accounting for perceived risk (Sjöberg and Herber, 2008). Additionally, he argues that epistemic trust should be distinguished from trust in experts because of the possibility that people may question science itself. According to **Sjöberg**[3], risk communication therefore, should use a more nuanced understanding of trust.

SECTION 2. SOCIETAL ISSUES

Basic and applied nanoscience and its interface with biotechnology and other sciences have challenged imaginations and for some evoked societal concerns. The term "societal implications of nanotechnology" includes an array of issues and is not limited to EHS issues. Non-EHS concerns include ethical, legal, and societal implications (ELSI) including privacy, legal issues, employment, economic development, access to education and job training, and infrastructure support and requirements, to name a few. For some publics, these concerns may be as important as, or outweigh, EHS issues. However, while ELSI issues involve important topics to many people, how these subjects are understood and addressed can be particularly difficult to quantify or measure. For example, discussing how nanotechnology affects privacy is highly problematic since limited data exists about the incremental effects of nanotechnology on privacy. In addition, efforts to design a calculus that enables comparative assessments of risks and benefits—given the general non-quantifiable nature of some values, like privacy—are evasive. Finally, a privacy concern to one group may be inconsequential or inevitable to another.

Discussions about net benefits depend on the intended audience and become exceedingly problematic when weighing a benefit of one nanotechnology development that may result in a cost to another. This underscores the need to disaggregate "society" or "the public" when thinking about risk communication...there is no "universal" audience.

EHS concerns are at the forefront of societal issues that have the potential to disrupt nanotechnology research and development (R & D) and applications. EHS communication about the potential toxicology of nanoparticles including inhalation, transdermal, and other translocation phenomena can be traced to a 2005 review article by Oberdörster, Oberdörster, and Oberdörster. Titled "Nanotoxicology: An Emerging Discipline Evolving from Studies of Ultrafine Particles," they argue that nanoparticles may require a toxicological methodology of their own because of their unique and unusual properties.

Some experts claim there is a difference between naturally occurring, incidental— often called ultrafine—particles or nanoparticles (NPs) produced by automobiles, incinerators, and fires, and engineered nanoparticles (ENPs). The differences between NPs and ENPs include: ENPs often have new and unique properties not observed in NPs. ENPs are mainly produced through controlled chemical reactions while NPs are often generated by the natural erosion and chemical degradation of paints, clays, etc. While people have evolved to live with many NPs, this may not be the case with ENPs. As Maynard (2008) noted, "just as the health risks from asbestos, vehicle emissions, and welding fumes differ, we will not be able to derive everything we need to know about ENPs just by looking at the incidental varieties."

Due to the impossibility of standardizing and evaluating every engineered nanomaterial, divergent viewpoints on how to proceed have developed. The *precautionists* want regulatory standards that limit entry of nanoparticles into the marketplace until safety, a topic with variable definitions, has been demonstrated. In contrast, the *cautionists* are developing a catalogue of characteristics of nanoparticles to allow regulators and researchers to assess, categorize, and rank-order profiles of nanoparticles. For example, some nanoparticles with high negative-risk profiles (very dangerous) could be treated differently than those with lower negative risk profiles.

Despite the consensus of technical experts that a nano-sized version of a substance may be different from a micro-sized version of the same substance, some are calling for a regulatory process that allows some nanoparticles onto the market if their non-nano- or micro-versions have already passed toxicological testing. However, engineered nanoparticles are being developed because they impart new properties. As such, new toxicological data on a nano-sized form of a substance may result in a different hazard assessment altogether. For example, nano-size particles and not larger particles may be able to enter the central nervous system via the axons of sensory neurons in the upper respiratory tract.

The list of characteristics of nanoparticles that may affect biokinetic risk profiles includes:

- chemical composition and contamination,
- number of particles per unit mass,
- median size and size distribution,
- physical form or shape,
- defect density,
- surface properties (area, porosity, charge, surface modifications, weathering of coating, etc.),
- agglomeration state,
- solubility and miscibility, and
- biopersistence.

Overall, the consensus of technical experts is that the lack of toxicological data on engineered nanoparticles complicates informed risk assessment and makes developing regulations especially challenging. An associated concern of the regulatory and environmental health and safety community is that industry, in its rush to get products onto the market, may simply exceed the capacity of the toxicological community to test nanoparticles.

The research needs assessment community has been behind the pace of product entry and continues to search for guidelines on what nanoparticles to study and how to study them best. While there are discussions among regulators about developing standards and guidelines, researchers are developing models that may assist stakeholders in tackling EHS issues of nanoparticles through application landscapes and exposure scenarios.

Recently, the EHS debate over nanoparticles has extended to the utility of lifecycle analysis. A lifecycle encompasses all the processes and activities that occur from the initial extraction of the material (or its precursors) from the Earth to the point at which any of the material's residuals are returned to the environment (Nano Risk Framework, 2007). This includes the cycle from product development to its disposal as waste.

Unfortunately, there are currently no standardized methods to complete a lifecycle analysis of nanoparticles due to problems in measurement, characterization, and pharmacokinetics. Nonetheless, coordinated efforts are underway.

Toxicologists often report their findings with appeals to uncertainty. Scientists can never predict risks with 100 percent accuracy. In many cases, identifying all potential risks may be impossible and scientists may simply be unaware of a potential risk (Smithson, 1989). Later discovery of a risk also can affect the public's trust regarding scientific assurances (see section 5). While incremental research findings may be reasonable for those familiar with science and science discovery, to the public, these shifts in information may be viewed as incredulous.

While the public, in general, may be accustomed to some uncertainty in their lives (i.e., weather, sports outcomes, political activity, etc.), they are generally more suspicious of scientific uncertainty, especially in areas of new science. For example, debates have been fueled over ozone layer mitigating chlorofluorocarbons and human- based carbon-dioxide-production-instigated climate change because of the public's observations of scientists disagreeing on the amount of risk the public will face. The uncertainty rather than the science becomes the focus in the media and the public (Friedman, Dunwoody, & Rogers, 1999).

Toxicologists also have used metaphors to shape imagery when communicating about risk. For example, in a recent report, asbestos and mesothelioma were used as a reference in a research project involving carbon nanotubes and mesothelioma (Poland et al., 2008). How should general readers understand the science of carbon nanotubes and mesothelioma when repeated inferences and references to asbestos rather carbon nanotubes were made throughout the article abstract and the discussion of the experiment? In this instance, asbestos and not the science of the study functioned as an established image reference and a vehicle for rhetorical fanfare.

The lesson to be learned: there are many voices that must be heard when communicating societal implications to the public. Science, and specifically toxicology, represents only a small number of voices. As we will learn below, the seemingly objective language of science does not translate clearly to a public audience.

SECTION 3. SOURCES AND MEDIA

Though some media theorists have argued that digital media will replace traditional media, media scholars such as Jenkins (2006) have argued that digital media will interact with and not replace traditional media in the immediate future. Already, a sizable minority of Americans get their news from both traditional media sources as well as the Internet with the proportion of Americans getting information from television decreasing while Internet sources continue to increase for nearly all demographics (Pew, 2008). Recently, data suggests television and Internet sources rank first and second far ahead of newspapers though much of what is found on the Internet is drawn from traditional media resources.

In risk communication, there are primarily four sources of information (see below) that have to be filtered through the media in order to reach the public. Integral in this risk communication process is the way the communication is framed by the original source and then translated by the media. This section discusses how risk messages originate, how they are dispersed, and how the public interacts with traditional and digital media. This interaction, in turn, can influence how messages are originally formed.

3.1. Sources, Amplification, and Attenuation

Risk-specific information on emerging technologies can be traced to four predominate sources:

- *Industry, marketing associations, and the insurance industry* publish reports on investment and risk. These reports tend to be technical in nature and seldom address risk as perceived by the public.

- *Government reports* have more regularly included discussions on a range of societal issues. In addition to technical risk assessment, these sources identify societal dynamics and ethical, social, and legal issues. These reports also address some public engagement efforts.
- *Academic (non-industrial) scientists* are a source of information on EHS and their material appears on-line, in professional technical journals, and occasionally in media reconstitutions of the findings, conclusions, and sometimes implications of their work.
- *Non-governmental organizations and civil advocacy groups,* including the Canadian ETC (Erosion, Technology, and Concentration) Group and the Australian branch of the Friends of the Earth have published statements and recommendations. For example, in 2004, the Action Group on Erosion, Technology and Concentration (ETC group) was the first advocacy group calling for a moratorium on the manufacture of synthetic nanoparticles created in the absence of **EHS impact analysis**.[4] More tempered voices have included the Environmental Defense Fund, the Natural Resources Defense Council, Consumers Union, and others.

3.2. Social Amplification and Attenuation

The Social Amplification of Risk Framework (SARF) (Kasperson et al., 1988) describes how social and individual factors amplify or attenuate risk perceptions. Kasperson modified this model (Kasperson et al., 2001) to accommodate stigmatization processes (see Figure 2). Helping to understand the "non-rational" ways people assess risk, SARF extends understanding of the effects of heuristics (mental short-cuts) and biases by explaining how heuristics and biases influence human perception (see section 4). In particular, an important contribution of the theory is the incorporation of "secondary effects" such as stigmatization of technologies, economic losses, or regulatory impacts (e.g., new airline security measures following September 11) to understanding risk perceptions.

According to the SARF, psychological, social, and institutional factors influence the public's risk perceptions and behavior through a network of socially mediated communication channels. These communication channels can be formal, such as the media, public relations campaigns, religious teachings, and community meetings, or informal, such as word-of-mouth interaction within social networks. While many variables can amplify and attenuate risk messages, researchers have focused on media, as the media have traditionally had a strong influence on risk perceptions (Pidgeon, Kasperson, & Slovic, 2003; Flynn, Slovic, & Kunreuther, 2001; Kasperson et al., 1988). However, in local risk debates, less formal sources like citizens' or activist groups can act as amplification or attenuation stations (Masuda & Garvin, 2006).

SARF details four main mechanisms of risk message amplification: signal value, stigmatization, heuristics and values, and social group relationship. Signal value refers to the idea that some events have an associated warning alarm for society. An event provides new information about the probability that a similar event will occur. It is considered high in signal value if it changes the perceived probability of future occurrences by the public.

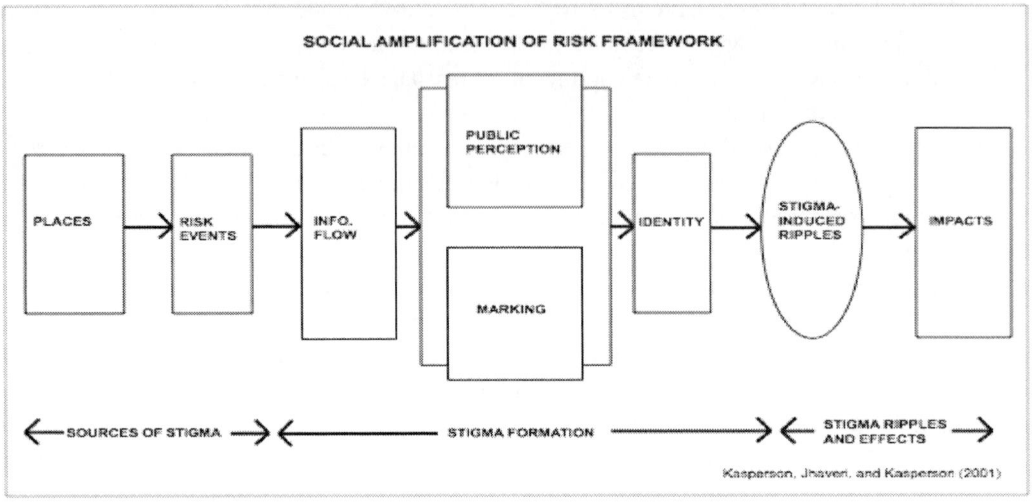

Figure 2. The Social Amplification of Risk Framework.

Stigmatization is the increase in the negative perceptions of an event (which could be an activity, person, place, or technology) that results in avoidance. Heuristics and values comprise the everyday values people hold that shape their responses to and decision-making behaviors regarding risk. In order to amplify a risk message, communicators use content that triggers attention and reaction. This information must be audience-specific (i.e., unambiguously understood by its audience as definitive and easily interpretable). An example of social amplification was the "mad cow" disease scare in the United Kingdom (Leiss & Powell, 1997). An example of inappropriate attenuation includes radon gas, automobile accidents, and smoking (Breakwell, 2000). Once these messages are sent, the social processes of amplification and attenuation inhibit the control of the message by the sender.

3.2.1. Amplification and Media: New Technologies Amplify Risk

Studies of traditional media (newspapers, magazines, and television) conclude that the general pattern for full articles is to provide reassurance and de-emphasize the severity of risks. Conversely, television coverage of environmental biotechnology focuses on extreme examples and unknown risks and is often superficial and incomplete. Mazur (1990) found that "extensive reporting of a controversial technological or environmental project not only arouses public attention but pushes it toward opposition" (p. 295). Renn (1991) also has argued that public attention or pure volume effect is only one of the numerous influences the media have on public perceptions of hazard. He also cites other effects used to influence public perceptions of hazards such as deleting and adding information, mixing effects (changing the order of information in messages), equalizing effects (changing the context), and "stereo effects" (multi-channel).

Media and amplification studies also have shown that only a few national news organizations are influential in selecting which hazards receive the most attention; public action to alleviate a hazard rises and falls with the amount of media reporting. However, with changes in digital media, the conclusions drawn from traditional media research may not be as predictive in the future.

3.2.2. Industry, Government, Non-Industrial Scientists, and Non-Governmental Organizations

Amplification can also occur outside the media. The concept of contagion is one most people can understand and relate to in their personal lives. In this subject field, contagion refers to the phenomenon where a negative event is inappropriately overgeneralized to other similar ideas, institutions, or events. The recent financial struggle of some mortgage companies provides a good example of contagion. Because one or few companies failed to fulfill their financial obligations, the public perceived that the entire industry was in crisis and made decisions accordingly.

As another example, spokespeople from the Pew Charitable Trust-supported Woodrow Wilson Center for International Scholars Project on Emerging Nanotechnologies (PEN) often use claims about contagion when addressing the public in person or in print about nanotechnology. David Rejeski of PEN has suggested that a contagion event will hurt the nanotech industry. He has argued that, "the U.S. regulatory policy for addressing nanomaterials so far has been ad hoc and incremental." He has expressed concern that until large-scale benefits of nanotechnology materialize— something that could take as long as a decade— "there will be little public tolerance of oversight failures or mishaps." Should a mishap occur, he said, it "could rapidly chill investments and galvanize public opposition" (Morrissey, 2004). Additionally, when speaking about a sealant spray that had been misidentified as including nanoparticles he said, "This really raises a bunch of interesting questions, since the public has been told that nano will cure diseases, not cause them. I think this is an important event in the nano world" (Weiss 2006, p. A02). Yet, two years after this "mishap" there has been no evidence of a contagion event. In 2009, a newspaper report (Lyn, 2009) reporting findings (Song et al, 2009) alleged the deaths of two Chinese women were linked to workplace exposure to nanoparticles and once again there was no evidence of a contagion event. For that matter, there is little evidence, so far, that contagion crosses product lines and/or industries. As such, a heterogeneous industry like nanotechnology seems less likely to experience a contagion event.

The food industry appears concerned about a "contagion effect" of nanoparticles in food suffering a similar fate in public perception as genetically modified (GM) foods. This is a serious apprehension given the losses in market capitalization associated with the GM debacle. Shares in the top firms dropped 43 percent between January and October in 2002 (Ho & Ching, 2002). However, researchers are beginning to show that the public may be more willing to accept nanotechnology food than GM food because most nanoparticles in food are actually of natural origin (Siegrist, 2008).

Though the media have the greatest influence on the amplification or attenuation of a hazard, industry, government and NGOs also influence media reporting. For example, as information sources on hazards, these types of aforementioned institutions may attempt to unduly influence the information and messages they provide to the media. Furthermore, the media may base decisions on what hazards to report based on recommendations from these institutions.

Because policy makers often sense public opinion through personal observations gleaned from the media rather than public polling data, NGOs can be very successful in influencing policy (Bakir, 2006). For example, NGOs such as Greenpeace, World Wildlife Foundation (WWF) and Friends of the Earth often amplify rather than attenuate hazards by emphasizing commercial product risks to consumer and citizen audiences. These advocacy groups recognize more media coverage can be attained by focusing on consumer products (Illes, 2007).

On the other hand, government and industry tend to attenuate risk perception more than NGOs. These institutions often attempt to trigger attention and provide information through public meetings. Because of the variety of public meeting contents, contexts, and attendance, these meetings do not uniformly amplify or attenuate risk perception or credibility ratings. Despite much recent criticism, it is important to note that research studies have shown public meetings generally enhance government or industry credibility (Wilsdon & Willis, 2004) and amplify unease (Cobb & Hamlett, 2008).

3.2.3. Mediation

The efficacy of message creation relies on the networks and the events by which these messages are formed. Potential network effects include an audience's personal experience and relationships, prior knowledge and education, cumulative influence of previous media reports and popular representations. Time, place, event, and other factors such as authority and situation are also network factors. Because many risk profiles of nanotechnologies and other emergent sciences are relatively unknown, numerous media (traditional and new) are beginning to blur the lines between fact and fiction in their reporting and programming. Here, messages are not simply amplified but are actively created—a perplexing problem for traditional sender-receiver/encoding- decoding models of information transmission.

Further complicating network effects, some public affairs departments are including information within television dramas, docudramas, and science shows, creating a new trend in fictional "reality" programming. Various dramas, movies, celebrity endorsements, and television-based news shows have created public concern and in turn a call for political action. For example, public affairs specialists at The John Hopkins Media Institution have persuaded shows like Chicago Hope and ER to include pre-produced consumer health messages (Rodgers, 1999). Whether these efforts have been effective forms of public health education have yet to be answered.

These entertainment strategies may be responsive to gender gap news interest preferences. Of those who follow science and technology news, 71 percent are men compared to 29 percent who are women. By contrast, of those who follow health news closely, 64 percent are women while 36 percent are men. Women also follow entertainment, celebrity, culture, travel, and community news more closely than men (Pew, 2008). Presenting science and technology information via entertainment popular with females for instance, may have a greater likelihood of reaching that population.

However, along with these entertainment strategies have arisen accusations of exaggeration and sensationalism (Weingart, Engels, & Pansegrau, 2000). For example, there is a fairly obscure 2004 Science Channel docudrama about Hendrik Schön of Bell Laboratories, who made extravagant and some unsupportable claims in nanoelectronics. In the docudrama, John Alexander, who has done some consulting with CINC US Special Operations Command, discusses **omnivorous nanocomputers**[5] (gray goo), a subject debunked by most nanoscientists. Of course, if the public discovers that a docudrama has been largely sensationalized after they have been made to feel alarmed, their feelings of being duped and betrayed may result in them rejecting risk messages because they no longer trust the source (see section 5 for more about trust).

Constructing the illusion of real science within fictional shows can confuse the public from distinguishing between what is fictional and what actually occurs in real life. An example of this blurring has been named the CSI Effect. In this case, the increase of interest in forensic science has resulted in desensitized and misinformed juries who expect the latest in TV-inspired DNA evidence and forensics at every trial, a public expecting a 40-minute investigation turnaround, and, ironically, more and better- educated criminals (Mertens, 2006). To date, there has not been a sufficiently important positive or negative fictional depiction of nanotechnology to generate testable levels of interest and attention, *Prey* and *G.I Joe* notwithstanding.

3.2.4. Opinion Leaders: Amplifiers and Attenuators

Although traditional forms of civic engagement have been in decline for the past several decades (Putnam,1995; 2000), opinion leaders, a group of people who actively participate in their communities, have served as amplifiers and attenuators. Scheufele & Shah (2000) have demonstrated that strong and persuasive personalities are related to effectiveness of opinion leaders. Opinion leaders with these personality traits can influence citizens' reactions to social issues because they are well-situated, well- prepared and better informed about current events, political matters, and other issues affecting their particular interests.

The concept of opinion leaders was first introduced into communication scholarship by Lazarsfeld, Berelson, & Gaudet (1944). They defined opinion leaders as "people who are most concerned about the issue as well as most articulate about it" (p. 49) and claimed they could be identified and studied by asking people whom they turn to for advice on any particular issue. They often have more exposure in media such as the radio, newspapers, and magazines and use the Internet to disseminate messages. Because opinion leaders are believed to have access to a great deal of potentially relevant information and people via voluntary associations, meetings, and other kinds of societal events, opinion leaders often operate as a nexus within specific human and material networks.

While it is important to provide citizens with information in order to encourage informed civic engagement, it is equally important to identify and direct credible resources to opinion leaders. Opinion leaders are information seekers who strategically use communication for civic participation. What opinion leaders consider important will be discussed by others and what they disregard will have a lower probability of being on the public radar. In the case of nanoscience, opinion leaders must be taken into account and provided with appropriate information. If opinion leaders are provided with accurate information, there is a greater likelihood that the public will be well informed.

3.3. The Role of Traditional Media: Priming, Framing, Agenda Setting

Recent research about the effects of mass media on lay audiences has been dominated by cognitive models of media effects including priming, framing, and agenda setting (e.g., Scheufele & Tewksbury, 2007; Scheufele, 2000). These models mark a sharp departure from traditional stimulus-response models that have historically shaped thinking about how communication about science influences audiences. Within this context, no assumptions exist about simple linear effects of informational messages. Rather model communication is viewed as an interaction between audience values and predispositions, on one hand, and mediated messages, on the other hand. They are often referred to as "negotiation models" of mass communication (McQuail, 2005).

3.3.1. Setting the Public Agenda

A fundamental function of mass communication is to set the public agenda. People generally do not maintain a running tally of all relevant political issues at a given moment. Rather, their thinking is dominated by a handful of salient issues. Why do certain issues become part of public discourse while others do not? The appearance of issues on the public agenda is usually a function of two related processes—agenda building and agenda setting.

Agenda building broadly refers to the idea that finite news spaces also limit the number of issues included on the public agenda at any given time (Cobb & Elder, 1971). The factors that influence which issues are included on the agenda and which ones are not can be grouped into three broad categories:

- *Factors intrinsic to the media system.* Shoemaker & Reese (1996) have reported that organizational structures, media ownership, news production norms, and ideological leanings of journalists influence how editors and journalists select issues (White, 1950).
- *Specific characteristics of news items.* Qualities or "news values" that influence journalists to select or adopt news items into public discourse include cultural or regional proximity of an issue to a target audience, unexpectedness, timeliness, and potential for negative consequences (Galtung & Ruge, 1965).
- *Factors extrinsic to the media system.* Efforts by political actors, interest groups, and other players in the policy arena can thrust specific issues onto the media and/or public agenda through press releases, paid communications, personal connections with journalists, and more recently online media.

Within the context of online media, the influence of blogs on traditional news coverage is noteworthy. Recent examples of blogs influencing public discourse have operated in a two-step flow model, where blogs either kept an issue on the public agenda by forcing mass media to continue to pay attention (e.g., Dan Rather's coverage of what turned out to be forged documents related to Bush's military service) or by introducing an issue onto the traditional media agenda that traditional media initially withheld (e.g., Matt Drudge's initial post on Newsweek "withholding" the Monica Lewinsky story).

Agenda building: Processes or efforts that determine which issues are covered by mass media.

Agenda setting: The link between the prominence with which an issue is covered in mass media and the importance attributed to the issue by the public.

Priming: A small set of the most salient issues in voters' minds often shapes their views on policy options

Agenda setting and priming are outcomes of agenda building. Agenda setting refers to the amount or prominence of coverage devoted to an issue (which is a direct outcome of the agenda building processes discussed earlier) that will influence the importance attributed to this issue by audiences (Scheufele, Shanahan, & Kim, 2002).

Salience is particularly important since it affects how easily accessible an issue is encoded into memory and, when needed, how easily the information can be retrieved to form opinions and judgments. Social psychologists often refer to the idea of memory- based models of information processing to explain salience (Hastie & Park, 1986). These models assume that attitudes are not stable but rather influenced by the considerations or issues that are most salient when citizens are asked to express an attitude (Zaller & Feldman, 1992). As a result, issues that are saliently portrayed by mass media are more likely to also be perceived by audiences as the most important. Salient issues will—in turn—influence judgments about policy options and candidates (Iyengar & Kinder, 1987). For example, during George H. W. Bush's re-election bid in 1992, when the mass media shifted their attention from the first Gulf War to a potential recession in the U.S., this shift effectively changed the standards that voters used to judge Bush and Clinton toward the end of the race.

"What" is being covered and "how" it is being covered is important in communication efforts. Agenda-building, agenda-setting, and priming are all based on the premise that the mere presence or absence of an issue on the public agenda can have significant impact on how people think about potential policy options or form other attitudes, regardless of the tone or slant of the message itself. However, the focus on what is being covered must be understood in terms of how issues are being framed and how framing and (re)framing can shape public perceptions.

3.3.2. Framing

> ***Framing:*** *The idea that the same piece of information can be presented in different ways using different terminology or visuals, and that these different presentations can produce very different interpretations of the same information.*

Salience is often created through discursive "framing." Framing is when the same piece of information can be presented in different ways that can alter how audiences perceive and think about topics or issues (Scheufele, 1999). The mechanisms behind framing are based on Goffman's (1974) work and Daniel Kahneman's Nobel Prize winning research on framing and how individuals process new information (Kahneman, 2003). Kahneman's work showed how "framing" or presenting a given piece of information will influence which interpretive frameworks audiences use when making sense of this information.

Framing, Kahneman showed, works best for "ambiguous" stimuli, (i.e., information that is not clear-cut and can be interpreted in different ways). For many citizens, science is such an ambiguous stimulus. Stem cell research, for instance, can be thought about as a moral issue, as a medical breakthrough, in relation to a specific celebrity, or as an issue with economic implications. Therefore, by framing a news story about stem cell research in the U.S. in moral rather than economic terms, for instance, mass media can indirectly influence people's interpretation of the issue.

Framing effects can be understood at social and individual levels. At the social level, framing is an important tool that is often used for processing information. For example, analogies and comparisons are used in order to explain issues to others and show their connections with what they already know. Similarly, journalists use frames to make relatively abstract political or scientific issues accessible to audiences. Additionally, framing is an integral part of news coverage. Journalists are often expected to reduce a complex issue to meet space restrictions such as an 800-word article or a 30-second segment on a local newscast.

Special interest groups, political actors, and other stakeholders in the policy arena frame issues in ways that will lead to more positive evaluations of their perspectives by voters, journalists, or editors. The terminology battles over the "death tax" versus the "estate tax" or the attempts by the Intelligent Design movement to reframe the issue of teaching evolution as an effort to "teach the controversy" are some recent examples. Greenpeace's campaign around the "Frankenfood" label (Nisbet & Scheufele, 2007) was a successful effort that established a frame early on in the issue cycle and dominated public discourse as a result. For nanotechnology, similar analogies and frames emerge. Is nano, for example, the next plastic or the next asbestos?

Here, an individual-level understanding of framing effects is critical. Why does a simple change in wording or labeling change the way we think about an issue? In order to make sense of the huge amounts of new information we must process on a constant basis, we quickly categorize that information using long-term schemas and value systems. Effective frames therefore use terminology or visuals that resonate with a particular schema and make it more likely that we interpret new information consistent with this schema (Scheufele, 2006).

Understanding the power of framing is critical for at least three reasons. First, all forms of communication are framed in one way or another. There is no frame-free communication. This includes news articles and most of our daily conversations. Paying attention to how messages are presented is necessary. Second, thinking carefully about our terminological choices is especially important for science-related issues. As the "Frankenfood" example shows, once frames are established in public discourse, they are difficult to change. Strategically countering misleading labels or frames early on in the issue cycle is imperative for any successful communication effort. Finally, the power of framing—in part—lies in the unconscious nature of its effects. In most cases, audiences find it difficult to distinguish the content of the message from the frame. While they may be able to selectively attend to certain messages, they will find it much more difficult to detect underlying frames.

3.3.3. Different Effects on Different Publics: The Role of Perceptual Filters and Values

Frames do not operate in a vacuum. For example, attitudes toward stem cell research show how frames interact with personal values where carefully researched "framing campaigns" on both sides of the issue have used people's religious beliefs and values. Recent studies (e.g., Ho, Brossard, & Scheufele, 2008) showed that ways scientific information reaches audiences and influences their attitudes differs depending on people's value systems or beliefs. In other words, frames may have very different effects for different audiences. Ho and her colleagues (Ho et al., 2008) demonstrated that levels of religiosity can determine if citizens use all the information in their possession about science when judging issues like stem cell research. In other words, religiosity served as a perceptual filter that determined which information people used and ignored—and held across issues, including nanotechnology (Scheufele & Brossard, 2008).

The idea of perceptual filters suggests that different groups may understand and react differently to the same information. Effective communication therefore needs to abandon "one size fits all" approaches that are sometimes used in science communication. Instead, communicating information will need to take into account the informational needs or concerns and the value systems of different audiences (for more on this see section 4).

This does not mean that an inherent disconnection between scientific progress and personal values among different publics exists. Rather, because audiences use their own values to "filter" new information (i.e., increased learning about a new technology), new information may produce different effects for different social groups. It is critical to develop a systematic understanding of these processes and to utilize this understanding to foster more effective dialogues with *all* groups.

3.4. Digital Media: Participatory Differences from Traditional Media

Different from traditional media, digital media is centered on the convergence of electronically or digitally-driven, computer-based activities using the Internet or cable as the delivery platform. With digital media, the public not only consumes the media but now participates in all three acts consisting of consumption, production, and sharing. What astonishes most people committed to traditional structures of distribution and consumption is that when given the opportunity to produce and share, the public often enthusiastically takes up the offer (Shirky, 2008).

Digital media is driven by the publics' attraction to participate in all three acts—consumption, production, and sharing—and is not a passing fad. Because media is key to communicating risk to the public, leaders must begin developing communication strategies that capitalize on using digital media, especially since the public expects to do more than consume and desires opportunities for interaction now more than ever.

3.4.1. Digital Media and Amplification

As types of media change, the way and degree to which media are capable of amplifying risk messages may also change. While Baby Boomers receive much of their news from traditional sources, a growing percentage of the American population relies on other news sources. Pew (2008) reports roughly 8 in 10 people say they watch, read, or listen to some kind of news. Additionally, the proportion of people who use a traditional news source on any given day has declined from 90 percent in 1990 to 74 percent today. Though the Internet has a smaller reach for news dissemination than other traditional media, its population of news consumers is growing while most other traditional media are shrinking. Today, 29 percent of Americans get online news, up from 23 percent in 2006. Television use remains fairly unchanged and the most widely used source for news, with 57 percent of Americans tuning in (59 percent in 1998). By comparison, newspaper and radio use for accessing the news have both dropped 14 percent from ten years ago, with newspaper news readership down to 34 percent and radio down to 35. In another study by Burns (2006), when asked which sources of news they expected to rely on in the future, 52 percent said they would "primarily" or "mostly" trust traditional news sources over emerging sources, and 35 percent said they expect to confer "equal trust" on both types of news outlets. Thirteen percent (13%) said they would expect to put more trust in emerging sources.

Critics suggest information collaboratively created online cannot be as credible as reviewed printed sources. However, in terms of accuracy *Nature* has reported comparable accuracy of Wikipedia to The Encyclopedia Britannica for articles on scientific subjects (Giles, 2005).

While the volume of voices of digital media bloggers may be high, voluminous, and exaggerated at times, the combined voices of Wikipedia, blogs (written and video), online portals (a directory of articles, blogs, etc.) podcasts (audio and video), and YouTube are competing and affecting how news information is communicated to the public.

Blogs clearly exemplify how journalistic practice has shifted. The list of nano blogs (Table 1) is just a sample of the new citizen journalist reporting that more and more people create and consume. In 2005, six percent of the entire U.S. population had created a blog and 16 percent regularly were reading blogs. Recently, Pew (2008) reported that nearly a quarter of Americans (23 percent) reported they regularly or sometimes read blogs about politics or current events.

The Pew Internet and American Life Project conducted a number of telephone surveys of Internet use in the United States, before and after 9/11. Their findings suggest that half of Internet users experienced problems accessing the news sites they were trying to reach and general use fell during the days after the attack. Even so, many people turned to alternative news sites, international news sites, and government and NGO websites, such as those run by the Red Cross, for information (Anderson, 2003).

Television was reported as the primary news source and the Internet was reported as significant in its interactivity for people posting personal accounts, photographs, and video footage of the tragedy. Journalists reported participating in Internet chat rooms and online bulletin boards looking for eyewitness accounts (Langfield, 2002). Some speculate that the use of these sources for seeking information and fact finding during the 9/11 crisis will lead to their being used again in the future if crisis or risk situations are presented. Below are distribution portal lists (Table 2) and podcasts (Table 3) that could provide potential sources of information about nanoscience (and in some cases other science information) for people to use while fact-finding during a crisis or risk situation.

Table 1. Examples of nanoscience blogs

Blog Titles	Author
Blog~Nano: Nano Scale Materials and Nanotechnology http://nanoscale-materials-and-nanotechnolog.blogspot.com/	Rocky Rawstern
Blog@NanoVic: Australian Perspectives on Nanotechnologies http://blog.nanovic.com.au/	Multiple contributors
Environmental Defense Fund: Nanotechnology Notes http://environmentaldefenseblogs.org/nanotechnology/	Multiple contributors (Richard Denison)
Nanodot: The Original Nanotechnology Weblog http://foresight.org/nanodot/	Jim Lewis
Nanohype: Nanotechnology Implications and Interactions http://nanohype.blogspot.com/	David Berube
Nanopublic http://nanopublic.blogspot.com/	Dietram Scheufele
SAFENANO Community http://community.safenano.org/blogs/	Multiple contributors (Andrew Maynard)

Table 2. Examples of nanoscience distribution portals

Portal Titles and URLs
AZo Nano http://www.azonano.com/ **Center for Nanotechnology and Society** http://www.cns.ucsb.edu/news-events/11/ **International Council on Nanotechnology (ICON)** http://icon.rice.edu/ **Nanotechnology** http://www.nanotechnology.com/ **Nanotechnology Now** http://www.nanotech-now.com/ **Nanotechwire** http://nanotechwire.com/ **Nano Tsunami** http://www.voyle.net/ **NanoVIP** http://www.nanovip.com/ **Nanowerk** http://www.nanowerk.com/ **Nano World News** http://www.nsti.org/news/ **Project on Emerging Technologies** http://www.wilsoncenter.org/nano/ **SAFENANO Community** **http://www.safenano.org**

Table 3. Examples of nanoscience podcasts

Podcast Titles and URLs
ACS Nano http://pubs.acs.org/journals/ancac3/podcasts/index.html **AZoNano** http://www.azonano.com/nanotechnology-podcasts.asp *Only nanotechnology podcast available on iTunes **Nanotechnology Podcast Blogs** http://www.digitalpodcast.com/detail.php?id=2000 **NanoVic** http://www.nanovic.com.au/?a=nanosociety.podcast&p=253 **The First Nanotechnology Podcast** http://www.podcastalley.com/podcast_details.php?pod_id=3832 **Small Talk** http://www.exploratorium.edu/ti/podcasts/smalltalk.php **Trips to the Nanofrontier** http://penmedia.org/podcast/nano/Podcast/podcast.html

The Internet has created a revolution in information accessibility. In contrast to traditional media, information on the Internet is more than ephemeral events but rather layers of enduring information with thousands of sites for finite topics. According to Krimsy (2007) studies looking at the most helpful sources of risk information for the public found that the volume and scope of risk messages were reported more extensively on Google than in academic or traditional media portals. Whereas traditional media might begin by amplifying a message because messages are more easily recognized, digital media may provide a potentially powerful, though mostly untapped, opportunity to attenuate a message. Unfortunately, because information flows faster, covers more sources, and is more accessible on the Internet, it is harder, though not impossible, to determine whether a risk event will be amplified or attenuated.

Some speculate that the Internet is becoming the central nervous system of a healthy democracy—the greatest avenue for free speech, civic engagement, and economic growth. Over the past three decades, the number of companies that control the media has shrunk from more than fifty to fewer than ten media giants and the Internet is no different. Of the top twenty most popular Internet sites in the United States, Google Inc. and Microsoft Corporation each own three (rankings based on data from Alexa.com, Top Sites United States, 2007). Nonetheless, unlike traditional media outlets, the content of a large number of even these websites is created by their audiences.

3.4.2. Digital Media and Science

Surveys of scientists indicate that one of their roles is to communicate research findings to political decision makers and to the broader public. They use the media to communicate more often than not. However, many researchers, including Jasanoff (1993), comment "...the public has a distorted view of risk because the media portray science in an inaccurate way, with exaggerated accounts of uncertainty and conflict" (p. 123). Although the news media plays a dual role as both disseminator and evaluator of the quality and validity of scientific information to the public, they are not particularly successful at doing either task well. Instead, the media frequently report "gripping instances of misfortune, whether or not representative of the activities that give rise to them" (Sunstein, 2005, p. 125). These reports are presented "often without enough facts to alleviate the possible fears they cause" (Wahlberg & Sjöberg, 2000, p. 34). The complex motivation for amplification is in part propelled by economic self-interest and the drive for increased readership and viewership, especially given recent competition from digital media.

3.4.3. Future of Digital Media

With the increasing sophistication and use of the Internet, public participation in the production, consumption, and sharing of media has grown tremendously. Some people want to be involved and express their opinions on topics they believe will affect their lives (for better or for worse). As compared to traditional public meetings that are typically held during working hours for most people, the Internet (24/7) allows for convenient and interactive communication and for more people to weigh in on any given subject through message forums, wikis, blogs, and other sites for interactive communication.

What does this mean for the discourse of science? It means a more involved and easily organized public that can respond to information from scientists and communicators. The power of digital media to spread the word to organize on even short notice should not be underestimated.

The future of digital media is more an issue of application than as a new technology. As technology improves, more processes may be merged. For example, the word "telephone" is almost a misnomer when applied to this technology since mobile phones are not just telephones anymore but are more like miniature computers. In Japan, the cell phone, a *keitai*, translates to "something you can carry with you" and does not include the word phone Science experts will need to understand that when communicating about science and risk, they will be reaching a much broader audience almost instantaneously through these technologies, resulting in people discussing what they learn from the experts in nearly real time.

3.5. Information Overload

While benefits exist for easier information access, researchers have found information overload (also referred to as: information anxiety, infoglut, data smog, data clouds, analysis paralysis, information fatigue syndrome) caused by the Internet can impede judgment, increase stress and physical ill health, and decrease job satisfaction (Lewis, 1996). Information overload is caused by a sense that a person is losing control over his or her ability to comprehend the meaning of complex and/or large amounts of information. Information overload is found over a wide range of disciplines such as medicine, business, computing, information, and the social and behavioral sciences (Wurman, 1990).

Some organizations have been designing methods to decrease information overload using design tools for evaluating and rating the quality of websites (see Edmunds & Morris, 2000). These tools aim to guide site developers, filter content, and help consumers become discerning users of information. To address these issues, Xerox, IBM, Microsoft, and Intel among other technology leaders have formed the Information Overload Research Group, whose goal is to determine how to increase workplace productivity. Interruptions and time needed to return to a task cost the U.S. economy $650 billion per year (Xerox joins IORG…, 2008).

3.6. Summary and Conclusion

As a recent Pew study reports, digital media is steadily growing more pervasive (2008). While a majority of Americans receive their news from traditional media, these traditional media readers are decreasing in number. As new information is continuously made available to the public via the Internet news outlets, messages and ideas spread on the Internet in a viral fashion. The public then encounters these messages and ideas and in turn communicates and disseminates them through message boards, blogs, and instant messaging venues. Unlike traditional media, more members of the public take an active role in communicating information about empirical data and/or personal opinions.

The speed of digital media offers two possibilities for risk communication. On the one hand, dangerous or catastrophic events can be communicated rapidly to the public with the possibility of being exaggerated before the experts present their interpretation of the events or findings. On the other hand, risk communicators and experts can also use this same opportunity to quickly and effectively communicate to the public. To do this, risk communicators and experts will need to become better skilled with these technologies in order to preempt potential overreactions as the public is made aware of potential risks.

Because the majority of Americans currently receive at least a portion of their news through traditional media, risk communicators should utilize digital media as a complementary, not competitive, media strategy. Additionally, in all types of media, risk communicators will need to be thoughtful and deliberate in how they frame information when communicating messages. If messages are inaccurately decoded and disseminated by the public over the Internet, risk communicators should make proactive efforts to re-frame messages through various forms of traditional and digital media. Because many people no longer synchronize their news gathering to evening news casts, to reach the greatest number and varieties of the public as possible, risk communicators will also need to consider using more creative venues including docudramas, social networking sites, podcasts, blogs and Internet portals. Eventually, these non-traditional formats could become the norm for the public's news consumption.

SECTION 4. RISK PERCEPTION

Sources of risk information define the scope of the message that is transmitted to and accessed by the public. However, at times, the risk messages received from media and other sources do not singularly define individuals' responses to hazards, but rather broader group or societal level risk perceptions. Source information must be filtered by the public for salience, re-organized for interpretation, and restructured to develop appropriate responses. This complex cognitive process people use as they receive risk messages and make meaning of them requires understanding risk perception. (At this point, the term "inexpert" will be used when discussing how the public makes sense of risk assessments in comparison to "experts" or those trained in science and technology. The use of "public" such as in "the public sphere" and "public" as citizen and sphere will be used when cited sources use the term.)

4.1. From Risk Assessment to Risk Perception

Quantitative risk assessment and risk management are typically based on a collection of expert-derived experimental and epidemiological data concerning the probability of exposure and dosage levels to the hazard of concern. These strategies are based on assumptions that individuals will rationally respond to technical assessment data and develop logical management policies. However, not everyone responds to risk in rational ways. Why are there more people afraid to fly than drive, when data show automobile accidents harm and kill more people than plane crashes every year? Why do people continue to use tobacco products when experts have shown the relationship of these products to increased incidences of cancer? These examples illustrate the importance of understand the power of human perception. "[H]ard facts only go so far, then human judgment is needed to interpret the message and determine their relevance for the future" (Slovic, Fischoff, & Lichtenstein, 1980).

4.2. Risk as a Socio-Psychological Construct

Renn (1992) suggests that when considering risk perception, definitions of risk must account for:

- how the likelihood of a hazard is specified and measured,
- how undesirable outcomes are identified, and
- how individuals define their reality.

Risk likelihood is often technically assessed as frequency of occurrence of an event averaged over time. Undesirable outcomes entail some evaluation of the level of harm to humans or ecosystems. If risk is defined as an objective reality, then assumptions are made that individuals can use quantitative and probabilistic data to make informed and rational decisions concerning personal and social reactions to a hazard.

However, when considering public perception, risk is not conceptualized as a simple mathematical function of event likelihood and severity of harm. "[D]anger is real, risk is socially constructed" (Slovic, 1999, p. 689). Thus, risk is understood by various factors, many of them qualitative in nature. In fact, members of the general public may view (un)desirable events through a broader lens defined by individual and group constructed social norms. Because hazard likelihoods are informed and influenced by values, source persuasion, personal experience, and culturally-based judgments, the public may not always agree that the empirical evidence offered by experts (who may be using different sources of data) leads to valid conclusions about risk (Bradbury, 1989).

Sjöberg (2001b) has stressed the importance of "risk as a construct…about realities." Thus, technical risk data is a critical component in public decision-making to the degree this information is integrated into individuals', groups,' and society's constructed perception. For example, "if you are run over by a car, it is likely to feel very real, even if constructs are used to talk about such events" (p. 120). For risk communicators, being able to understand similarities and differences between the realities *and* perceptions of risk is needed for effective message construction.

4.3. The Risk Perception Gap between Experts and Inexperts

If risk perception is socially constructed, differential risk evaluations between experts and inexperts may exist. Researchers have shown that a divide exists between how most individuals assign importance to risks and the likelihood of the occurrence or severity of effects as determined by expert evaluators (Kraus, Malmfors, & Slovic, 1992; Slovic, 1987). Table 4 lists some hazard domains where researchers have shown perceptual gaps between experts and inexperts. Note that overall risk evaluation seems to be bi-directional, with experts not always perceiving higher risks than the general public.

Why is nuclear power viewed as more dangerous by inexperts? Why is the risk perception gap reversed for electric power? Why are many recreational activities (skiing, swimming, etc.) viewed as more risky by experts than by inexperts? One hypothesis is that experts and inexperts (the public) possess different epistemologies when understanding and evaluating hazards. It is typically assumed that expert scientific knowledge is objective, verifiable, derived methodologically, and peer- reviewed. On the other hand, public knowledge is regarded as more linear, speculative, and socially embedded (Petts & Brooks, 2006). If experts and inexperts are using different languages for communicating risk, this creates additional communication challenges. Misunderstandings and miscommunications could occur when different stakeholders do not share similar values about research data and risk perceptions. Additionally, experts could become frustrated when the public's reactions to regulatory measures do not align with what experts assess as appropriate risk reduction strategies. Reducing the perceptual gap and finding a common ground that allows for the intersection of the public and expert spheres is critical for effective communication efforts.

4.3.1. Defining the Expert Sphere

Obtaining a common understanding whereby communication efforts can be mounted requires several definitions. Who are the experts? How valid and reliable are the data they rely upon for their risk assessment? Who are inexperts? How homogeneous are inexperts? Answering these questions is critical to understanding the risk perception gap.

So, who is an expert? Does possessing large volumes of general knowledge necessarily qualify one as an expert? Does being a research chemist qualify one to be an expert in toxicology? It is important to realize that many of the studies listed in Table 4 define experts very broadly (see Rowe & Wright, 2001 for a review of many of the studies). Experts are people who have a large volume of experience and information within a specific contextual domain. Membership in these groups is reinforced by domain-specific socialization, conformity pressure, political ideology, credentialing, and institutional affiliation. Sjöberg (2006) suggests that many science and technology experts, if clearly defined within a specific context, rely more heavily on quantitative data to assess levels of risk.

Experts are not, however, isolated from the public sphere and are not necessarily exempt from the power of qualitative perceptive factors in risk evaluation. Sjöberg (2006) urges caution when assuming a purely rational assessment of risk by experts since they are also subject to influences that can affect their judgment and add value- related factors to their evaluation. Experts are parents and consumers too. As such, they are members of multiple public spheres and are also influenced by the contextual factors of those spheres.

4.3.2. Defining the Public Sphere

Defining the public sphere is difficult. Often misunderstood to be a homogenous group with negligible individual variability (in light of large sample sizes and quantitative averages), few messages are well-designed for a large, general public sphere, but more typically are aimed at targeted sub-populations. Sub-populations often possess different demographics and characteristics than a broader general public. Because of this, it is more appropriate to define the public sphere as a compilation of smaller sub-spheres.

If the numbers are sufficiently high, the next assumption might be that regionally confined sub-populations could be representative of the population at large. However, with the high levels of mobility in the United States, individuals are self-organizing into communities that are increasingly like-minded. Called the "big sort effect," Bishop has shown that these polarized units become exposed to smaller amounts of contradictory information (2008). As insular groups become more homogenous, their views become more extreme and this is amplified or attenuated through digital media. Geographically oriented, these communities can also form online around specific forms of media and have similar demographic characteristics. For instance, the cable shows *The Colbert Report* and *The Daily Show* have less than a quarter of their audience over the age of 50 (Pew, 2008).

4.3.3. The "Big Sort" effect

It has become less relevant to regard the general public as an amorphous mass of averaged opinions. The American landscape is undergoing a "big sort," with the general public forming communities of interest and opinions in regionally discrete areas. This effect is likely to be amplified as opinion nuclei become more extreme and attract more like-minded citizens (Bishop, 2008; Florida, 2008). Therefore, the discourse in risk communication needs to evolve from addressing the perception gap of the general public to addressing gaps of specific public nuclei and their opinion leaders (see chapter 6 for information on public engagement).

4.3.4. Interactions between Expert and Inexpert Perceptions

The interaction between expert and public risk perceptions can lead to dynamics of contrasting views. With the technical experts using more empirical assessment data (but not without subjective influences), members of the public are more likely to use a type of meta-judgment when making decisions concerning risk (Sjöberg, 2006). In this regard, the public will balance multi-dimensional, qualitative factors with expert assessments that in turn are balanced by public perceptions about the validity and reliability of particular experts.

Table 4. Differences in expert and inexpert evaluations of risk from various hazards

Experts Perceive Lower Risk
Chemical Products (Kraus et al., 1992; Slovic et al., 1995)
Radioactive Waste Disposal (Flynn et al., 1993; Sjoberg & Drottz-Sjoberg, 1994)
Nuclear Power (Slovic, 1987)
Police Work (Slovic, 1987)
Mountain Climbing (Slovic et al., 1980)
Warfare (Savadori et al., 1998)
Healthcare Efficiency (Savadori et al., 1998)
Spray Cans (Slovic, 1987)
Food Microbiology (Jensen et al., 2005)
Nanotechnology (Siegrist et al., 2007)
Expert/Inexpert Agreement
Oil & Gas Production (Wright et al., 2000)
Nuclear Technology (Purvis-Roberts et al., 2007)
Radon (Sjoberg & Drotz-Sjoberg, 1994)
Inexperts Perceive Lower Risk
Electric Power (Slovic, 1987)
Surgery (Slovic, 1987)
X-rays (Slovic, 1987)
Swimming (Slovic, 1987)
Downhill Skiing (Savadori et al., 1998)
Bicycles (Slovic, 1987)

If risk perception is built upon qualitative evaluations about risk, these factors must be identified and addressed. Numerous factors have been shown to be significant in public risk perception including, but not limited to (see Slovic, 1992 for a review):

- the voluntary nature of exposure,
- the affective fear associated with the hazard,
- trust in the managing institutions,
- unnaturalness of the application, and
- the novelty of the hazard.

4.4. Rationality in Risk Judgments and Decision Making

The use of qualitative factors to make risk judgments implies inexperts will use spurious information and act irrationally or unpredictably in response to a given hazard. This offers little reassurance for those designing targeted communication messages. Early quantitative experiments seem to confirm this assumption. As Kahneman and Tversky (1973) observed, "[i]n making prediction and judgments under uncertainty, people do not appear to follow the calculus of chance and statistical theory of prediction. Instead, they rely on a limited number of heuristics, which sometimes yield reasonable judgments. Sometimes, they can lead to severe and systematic errors (biases)" (p. 48).

Kahneman and Tversky's research in human rationality in judgment and decision-making yielded three important conclusions (Samuels & Stich, 2004):

- People's intuitive judgments on many reasoning and decision-making tasks do not fit the appropriate norms of rationality.
- When deviating from appropriate norms of rationality, individuals tend to rely on mental short-cuts (heuristics).
- The only cognitive tools available to in experts are heuristics.

Evolutionary psychologists (Samuels & Stich, 2004) have called into question some of the veracity of these claims (especially the final one). They argue that, depending on the context, individuals can make judgments approximating the norms of technical rationality. In addition, they argue that heuristics are not the only cognitive tools used in decision-making. However, in the case of science and technology, where information is limited or uncertain it is not unreasonable to think that the inexpert will revert to a reliance on some cognitive short cuts.

4.4.1. Heuristics and Biases

Heuristics are simple mental formulae that help individuals make decisions and form judgments (for a list of common heuristics see Table 5). They are functional methods for arriving at a solution that is a reasonable approximation of the best answer. In a world that is often overloaded with sensory stimuli, cognitive heuristics allow for efficient information processing without demanding absolute accuracy.

Table 5. Common heuristics used in decision making. These are mental short-cuts that allow individuals to arrive at answers that are a reasonable approximation of the truth

Cognitive Heuristic	Pertinent Citation	Mechanism for the Decision-Making Heuristic
Affect	Finucane, Alhakami, Slovic, & Johnson, 2000	Rapidly induced, emotional reactions to certain risk information can override judgments based on logical inference.
Anchoring/ Adjustment	Tversky & Kahneman, 1974	High value is placed on a single informational unit, and all further information is adjusted in reference to that anchor.
Availability	Tversky & Kahneman, 1974	Risk information or hazard events that are easily called to mind are viewed as more risky or more likely to occur than probable events.
Familiarity	Tversky & Kahneman, 1974	It is assumed that information that was true for past situations consistently holds true for novel, unfamiliar situations.
Minimax	Waller & Covello, 1984	Value is placed on decisions that reduce the maximum possible loss instead of most significant losses.
Peak-end	Kahneman, 1999	Past events are defined and remembered by the extreme high and low points; mid-range events tended to be cognitively discarded.
Represent-ativeness	Tversky & Kahneman, 1974	It is assumed that information collected from small, familiar samples is representative of the population at large.
Simulation	Kahneman & Tversky, 1982	Value is weighted on events that are more easily imagined visually in an individual's mind.
Social Proof	Cialdini, 1993	In situations where individuals do not know the proper social action, they will look to others to define appropriate behavior and values.
Take-the-Best	Gigerenzer & Golstein, 1996	When making judgments based on multiple criteria, the criteria are tried in series and the first criterion that discriminates is highly valued.

Using these heuristics can lead to polarized tendencies in decision-making that manifest as "errors" (biases) inconsistent with logic rules or probability. The differences between heuristics and biases are subtle—they work in concert, with one reinforcing the other to determine judgments. Table 6 is a partial list of common biases that manifest themselves in decision-making under uncertainty.

4.5. Heuristics and Nanotechnology Risk Perception

Social and behavioral science research in nanotechnology is an emerging field with limited epidemiological and empirical data to support an understanding of the risks that nanotechnology enabled applications may have for human societies. This requires experts and inexperts to address how they will deal with uncertain and incomplete data. This uncertainty can also translate to disagreement between experts about the severity of potential risks. For inexperts, this disagreement may be skeptically viewed and may likely result in decision-making processes that rely more on heuristics than on other more rational means. Because of this, heuristics and biases that may be important when communicating about nanotechnology are addressed below.

4.5.1. Affect Heuristic

One of the most important heuristics likely to influence nanotechnology risk perception is the affect heuristic (Kahan et al., 2008). Visceral and emotional reactions to new technologies often shape the judgment process and help define the breadth and limits of expert information. Affect also manifests itself in cognitive biases such as the alarmist bias (where the "worst-case" scenario is recalled more easily and seen as more probable). When information concerning nanotechnology is scarce, members of the general public may rely upon their affect heuristics to shape their judgments about risks and benefits.

Table 6. Common biases in decision making. These are decision-making "errors" that commonly occur due to the use of a cognitive heuristic

Decision-Making Bias	How the Bias Manifests Itself	Pertinent Citation(s)
Alarmist	"Worst case" risks are seen as more probable and more valid than less extreme or more probable hazard events	Viscusi, 1997
All-or-Nothing	The only acceptable risk is seen as zero-risk	Karetz, 1989
Confirmation	Once a belief is in place only information that "proves" current beliefs is assumed valid	McGrew, 2003
Contagion	Contaminated objects may carry or transfer their negative essence to other objects with which they come in contact	Scherer & Cho, 2003; Kaufman, 1994; Douglas, 1990
Effort	Information or events that require more effort to obtain or occur are attributed more value than those perceived as "easy"	Kruger, Wirtz, Von Boven, & Altermatt, 2004
Hindsight	The likelihood of the occurrence of a hazard event is judged higher when perceived in hindsight	Fischoff, 1975
Optimistic	Individuals are likely to assume that they are less at risk than the general public for a particular hazard	Price, Pentecost, & Voth, 2002
Proportionality	Proportions of groups are considered of more value than pure numbers of individuals exposed to a hazard	Baron, 1997; Kahneman & Tversky, 1982
Probability Neglect	Individuals ignore probability in light of other evidence	Sunstein, 2002
Recognition	Information that is recognizable is given more weight than information that is not recognized	Goldstein & Gigerenzer, 2002
Risk Aversion	Individuals are hesitant to accept a bargain with uncertain payoffs rather than a bargain with certain, but possibly lower payoffs	Yaari, 1987
Scarcity	High value is placed on objects that are more likely to be lost	Cialdani, 1987

4.5.2. Anchoring and Adjustment Heuristics

The anchoring and adjustment heuristic refers to individuals' use and reference to numbers as anchors (Kahneman and Tversky, 1973). For this heuristic, Kahneman and Tversky found that if individuals were anchored with a number at either high or low extremes (say fatality estimates) and then asked their perception of actual statistics, they adjusted their estimate based on the initial extreme anchor number.

For example, people were asked to estimate the actual age when Ghandi passed away after being asked either "Was Ghandi older or younger than nine when he died?" or "Was Ghandi older or younger than 140 when he died?" Those who were given the lower "anchor" number estimated a much younger age (50 years old) than those who heard the higher number first (67 years old).

This anchoring heuristic's effect persists even when study participants are told about the technical aspects of the anchor heuristic or that the anchor used was an extreme over- or under-estimate. Thus, appropriate anchoring of concepts will be critical in nanotechnology risk communication. Additionally, beyond numerical data, the initial perceptions that individuals ascertain will likely shape the future directions to which they adjust their levels of acceptance or rejection. For example, early nanotechnology product entrants into the consumer market judged positively could anchor perceptions and opinions, to a degree whereby subsequent negatively perceived entrants may be viewed less negatively (Berube, 2009).

4.5.3. Availability Heuristic

The availability heuristic is a mental short-cut where individuals base their judgment of a particular event based on how easily an example can be recalled. For example, if individuals are asked to make a list of words that end in "-ing" versus lists of words where the second to last letter is the letter "n", they are more likely to come up with several words for the former while struggling with the latter. Yet, the tasks are similar. Words ending in "ing" are more cognitively available than words where the second to last letter is the letter "_n_". Because of the availability heuristic, it may be important to communicate a science-supported referent rather than a fictional one such as self-assembled nanobots. Once the public envisions nanoscience as nanobots, it might be problematic for realistic depictions to compete with the imagined.

4.6. Biases and Nanotechnology Risk Perception

Various biases can potentially be manifested in the discourse about nanotechnology risk. The following is a list of how some biases may arise when making judgments or decisions concerning nanotechnology.

- *Alarmist bias*: The worst-case scenario surrounding nanotechnology will be assumed to be the most valid. Drexler and the rest of the community have spent twenty years shaking free of his "grey goo" metaphor (1986).
- *Confirmation bias*: Individuals whose established beliefs filter information that confirms or reinforces those beliefs. For example, nanobiotechnology-driven life

prolongation efforts can be seen as "playing god" or "being unnatural" and might be particularly sensitive to this type of information filtering.

- *Contagion bias*: Objects that are contaminated (including ecosystems and nanofabricated objects) may be seen as possessing all the negative properties of only a few potentially toxic nanoscale objects. Large negative toxicological studies might serve as icons for negative perceptions of all nanotechnology applications though the heterogeneous nature of the industry suggests otherwise.
- *Probability neglect*: The tendency is for individuals to ignore probabilities and instead focus on all-or-nothing estimates of risk often based on anecdotal evidence. The current limited availability of data may contribute to the neglect of probabilities in risk estimates of nanoparticles. However, probability neglect among inexperts is pervasive, making them particularly vulnerable to overclaims. For example, if some nanoparticles from a topically applied lotion get into the bloodstream via a break in the skin, even though this mode of transmission and the level of exposure may be insufficient to produce negative health consequences or the particular nanomaterial may be benign at almost any exposure level, inexperts may ignore probability and assume any exposure may be hazardous.
- *Recognition bias*: Information that is recognizable is given more weight than information that is unfamiliar. Recognition bias is closely associated with the availability heuristic discussed above. In the case of nanotechnology, much of the data are not familiar or recognizable to inexperts. Rhetoric and images that connect nanotechnology application risks to other more familiar risks will likely carry greater perceptual influence. We see this in the association that some have made of some nanoparticles with asbestos in terms of inhalation- related toxicologies.
- *Risk aversion*: A behavior where individuals are hesitant to act upon information with uncertain payoffs in contrast to making decisions with smaller but certain payoffs. Nanotechnology acceptance may not be influenced by potential benefits, because individuals may rely more on classic technologies with less uncertainty. This bias may negatively impact chemical substitution and pollution remediation strategies whereby a known risk to health may be retained due to uncertainty values associated with the nanoscience-related alternative.

4.7. Challenges to Nanotechnology Risk Communication

Nanotechnology is often perceived as a singular field, but in fact, consists of numerous and diverse fields and applications utilizing the emergent properties of particles and materials structured at the nanoscale (see section 2). In addition to the variety of applications, nanotechnology has generated divergent views concerning the potential risks resulting from widespread use of nanotechnology. In turn, this has lead to expert disagreement and uncertainty. Importantly, the unique characteristics materials have at the nanoscale when compared to the same materials on the micro- or macroscale will also create perceptual boundaries that are novel in the field of risk communication and will require a close consideration of risk as a social construct.

4.7.1. Public Knowledge of Nanotechnology

Issues in communication are compounded by the fact that the general public has little knowledge about nanotechnology (Lee, Scheufele, & Lewenstein, 2005; Cobb & Macoubrie, 2004; European Commission, 2001). An interesting observation: public information and understanding is not increasing over time even when tracked against increasing coverage and outreach efforts.

Studies from 2004 to the present have not demonstrated a significant increase in understanding about nanoscience, a subject that needs to be better understood.

Despite this lack of information, individuals in the United States have been reported as perceiving that the benefits of nanotechnology will outweigh the risks even when asked about nanotechnology in the abstract rather than about specific applications (Cobb & Macoubrie, 2004).

Future research and communication efforts should be directed towards specific nanotechnology applications while informing the public about the heterogeneity of the field. Current nanotechnology applications span from everyday consumer products to more complex medical technologies. It is likely that the directionality and magnitude of the perceptual gap will vary depending on the specific nanotechnology application. As such, a critical key to effective communication of the risk of heterogeneity of nanoscale sciences and nanotechnology applications may be message specificity, both in topic and target audience. The benefits of risk messages grounded in explicit applications versus nanotechnology as a whole has been demonstrated in reduced risk perceptions for biotechnology-based medical applications (for which more benefits are perceived) versus applications utilizing biotechnology for food-related purposes (Savadori et al., 2004).

4.7.2. Experts and Nanotechnology

Differences in risk perception between inexperts and experts exist for nanotechnology. In general, inexperts report significantly more concern for potential hazards associated with nanotechnology than experts. Nanotechnology experts assess risk based primarily on the confidence they have in the science they are familiar with and, more generally, government oversight. Some researchers have shown that inexperts seem to make their risk evaluations based on the trust they have for the communicating and managing institutions (see section 5.1), the perceived benefits of nanotechnology, and their attitude about technology in general (Siegrist, Keller, Kastenholz, Frey, & Wiek, 2007).

Narrowing the perceptual gap is not simply a function of giving inexperts more information (see section 7.1.1.) but will require better understanding how both experts and inexperts cognitively select and sort information. Thus, nanotechnology experts will need to be better stipulated and their specific domains of expertise better described. This strategy may alleviate some needless confusion that might occur when experts differ when analyzing similar data sets.

The limited quantitative data and uncertainty surrounding epidemiological and toxicological data will further require experts and their expertise to be defined and circumscribed to enhance public trust and information accuracy.

4.8. Summary and Conclusion

Applying risk perception research findings is critical in assessing and understanding how inexperts and experts alike react to risk messages. The perceptual gaps between the public sphere and the expert spheres often arise when risks are being assessed. The following lessons for risk communicators have emerged out of the study of risk perception:

- The public sphere assesses risks based on a variety of factors that can include quantitative technical risk assessment but often is weighted towards qualitative judgments based on cultural values.
- Differential perceptions of risk leading to perceptual gaps between experts and in experts must be taken into consideration when designing communication messages.
- Defining the public and expert spheres is becoming increasing complex as the nation self-organizes or sorts into geographically and informationally concentrated areas.
- Making decisions about risks based on perceptions that use mental heuristics can translate into biased judgments. These judgments may be considered "irrational" but are often consistent and predictable.

The challenge risk communicators face is developing messages about nanotechnology that are science-supported, and that account for risk perception drawbacks while informing the public as to the heterogeneity of the field.

It may be useful to map public responses to releases of nanotechnology- associated products that receive favorable responses. For example, if useful medical applications appear earlier than other applications, these examples may serve as a positive "anchor" about a class of nanotechnology-related products. In turn, if a positive contagion effect occurs, these results may influence public responses about downstream products, albeit positive or negative information about nanotechnology (Berube, 2009).

Trust is another factor that plays a major role in shaping the risk perception of the public in nanotechnology and emerging sciences. The role of trust cannot be overemphasized.

SECTION 5. TRUST, FEAR, AND BELIEF

As discussed in section 3, the sources that members of the public use when receiving risk information are continually becoming more diverse. The public's trust in larger institutions is declining, and is paralleled by declining trust in the traditional media. Even so, research on which sources are more highly trusted by the public consistently report that media are more highly trusted than the government. For example, BlogHer and Compass Partners, LLC researchers found that approximately half of the women interviewed (over 5,000) consider blogs a "highly reliable" or "very reliable" source of information about everything from products to presidential candidates. As BlogHer co-founder Elisa Camahort Page explained, "We are losing trust in the government, politicians, the media and many corporations. But as it turns out, we trust each other. The blogosphere is increasingly becoming the digital 'kitchen table'" (New BlogHer, 2008).

The public may trust these new forms of media more so than larger institutions due to the historical lack of transparency and immediacy from the larger institutions. In numerous events where the public was told a phenomenon was safe, people later reported feeling betrayed when updated science-based information was provided and it was revealed that information had been withheld. Leaders of institutions seem to trust scientific risk finding. Unfortunately, due to disagreements among researchers, purveyors of fear have used contradictions in scientific findings to forward their own agendas, resulting in the public losing epistemic trust (trust in accumulated systematic knowledge). The following sections discuss the interplay between trust and fear and how these factors largely determine whether or not the public accepts a risk message.

5.1. Trust-Based Arguments

In risk management, trust is one of the most important factors for inexperts, especially when it is difficult for an individual to avoid the source of the risk. In this situation, inexperts are left relying on government and industry leaders to consider public welfare as important when making decisions. Even so, the public cannot be simply told they are safe. Top-down communication is not effective, especially when the source of the message may not be trusted and when the subject is exotic, such as "invisible" nanoparticles. Only when trust has been established will inexperts begin to welcome messages from experts. Experts and risk communicators must recognize that cultivating and maintaining trust is an ongoing professional responsibility and not something that can be optimally initiated or "geared-up" at specific times, such as in a crisis. Importantly, scientists and government officials need to recognize that trust is earned, not assumed, and cannot be taken for granted.

5.1.1. Proclivity to Trust

An increasingly skeptical public is losing its propensity to trust experts. According to a National Opinion Research Center survey, trust in experts is steadily declining. A comparison between views expressed in the 1970s (1972 - 1982) and 2004 showed declining confidence in banks and financial institutions (35% →28%), major companies (26% →17%), the press (24% →9%), organized religion (35% →24%) and education (36% →27%) (Tyler, 2006).

Trust can also vary according to specific issues and general attitudes towards institutions. For example, in Europe, trust varies widely because of differences in national attitudes to government more so than to specific (bio) technology-related issues (Gaskell & Bauer, 2006). Additionally, Putnam (2000) refers to a "virtuous circle" of trust, or "social capital" that consists of group membership and informal social ties. He blames the decline of social capital on isolation resulting from technology, especially the television and Internet. However, Putnam's claims have not been supported in subsequent research (Ulsaner, 2004; Norris, 1996, 2000, 2001; Newton, 1999).

While the public might have less trust in institutions, trust in family and friends continues to exist. Additionally, personal relationships were found to be more trusted than digital media. The Canadian research firm POLLARA (2008) reported that self- described social media users put greater trust in friends and family online than in popular bloggers or strangers with 10,000 MySpace "friends."

At the same time, these results may demographically vary. For example, a study concerning GM food reported opposing trends among younger demographics. Younger groups indicated that they would be "willing to tolerate the availability of GM food if it were always clearly labeled, whereas middle and older groups would not" (Hunt & Frewer, 2001). Here, younger consumers reported greater trust and higher degrees of knowledge than middle and older age groups. This suggests the possibility of developing better trust relationships with individuals of younger demographics though data on this remains mixed and context specific.

5.1.2. Types of Trust

Cultivating and maintaining trust is harder to accomplish than prescribe because trust is a densely layered variable composed of different factors working together such as believability and reliability.

Slovic (1993) has claimed that trust is fragile and that once lost cannot be restored. Known as the asymmetry principal, negative events can anchor and undermine trust and can also overshadow good news. Slovic witnessed this asymmetry in his work about nuclear power. However, his critics have suggested that his observations about asymmetry may have been artificially inflated by his choice of subject and public perceptions about nuclear power. At the time of his study, nuclear power witnessed a unique series of events including Three Mile Island and Chernobyl that polarized and anchored public perceptions and attitudes on nuclear energy and nuclear safety. In contrast, in the case of overall government policies the asymmetry did not appear. In the case of nanotechnology, to date, there has not been a singular (substantive) event to function as an anchor.

Trust and regulation concerns can connote regulation itself. To this point, as perceived by inexpert populations, regulation by the government may suggest a large degree of risk and the government's distrust of a particular technology.

Trust has been measured as *specific trust* and *general trust*. Specific trust involves trust as a single case or phenomenon and agents responsible for solving the issue. General trust refers to a larger set of phenomena or a much bigger institution and adds little to the explanatory power of trust, while specific trust, Sjöberg (2001a) claims, is a more powerful construct than general trust for explaining risk perception because of its detail and explicitness.

Social trust, the trust discussed most often in the literature, has been shown to be a relatively weak factor when compared with epistemic trust. Social trust partly explains epistemic trust and is a factor during less important issues when inexpert populations use peripheral processing. Epistemic trust involves people's use of central processing as a factor in perceived risk. Central processing is the mental procedure an individual engages in when a message is highly relevant (salient), causing the person to become more involved. In this case, messages are analyzed more carefully and any arguments made are scrutinized to see if they make sense based on prior experience. In this way, central processing has much in common with the concept of salience described earlier. Peripheral processing is considered less salient and is affected by cues in the message such as the credibility of the speaker or the speaker's charisma and persuasiveness (Petty & Cacioppo, 1986).

In fact, some failures of risk communication have been attributed to too much reliance on social trust, the trust people have in an institution based on their perception of the institution's goals, motives, and actions in relationship to their personal values (Earle & Cvetkovich, 1995).

It is important to recognize that trust in experts is a form of social trust. Researchers have shown that a weak relationship exists between social trust, risk perception, and the acceptance of hazardous technologies and facilities (Sjöberg & Herber, 2008; Sjöberg, 2008). For example, Sjöberg (1999, 2001a, 2004, 2008) has shown that people who say that they trust experts and organizations may not accept the policy these experts or organizations recommend because they may not be convinced about the trustworthiness of the scientific knowledge being used. The history of technology research, where hazards have occurred after assurances of minimal risks to the public (i.e. Chernobyl and Bovine Spongiform Encephalopathy [BSE]), could account for people's suspicions that the effects of technology are not yet understood even if they perceive the experts as honest and competent. This is because they feel experts' recommendations may be changed in the future when new scientific findings are reported.

5.1.3. How the Public Determines Trust

The public relies on trust, especially when faced with making a decision and they lack information or a predisposed opinion regarding risk hazards (Jenkins-Smith & Silva, 1998). However, there are many factors involved in how this trust is determined. Some trust models insist that empathy—perceived value similarity and heuristics—is a greater determinant of trust than carefully reasoned arguments or direct knowledge (Bakir, 2006; Poortinga & Pidgeon, 2003, 2005, 2006; Earle & Cvetkovich, 1995). For these models, sources of risk information must appear to care about the public and be knowledgeable on a specific subject. In cases associated with food-related hazards, degree of knowledge and concern were identified as important determinants of trust (Rosati & Saba, 2004; Hunt & Frewer, 2001; Peters, Covello, & McCallum, 1997). These and other factors including care, concern, values, knowledge, competency, dedication, and commitment fall within one of Poortinga & Pidgeon's two components of trust—general trustworthiness.

Additionally, vested interest is important and involves the (lack of) integrity, cynicism or reporting bias (Hunt & Frewer, 2001; Frewer, Howard, Hedderly, & Shepherd, 1996). Sources perceived as protecting their own vested interests are much less likely to be trusted. In this way, the strong American value, the pursuit of self-interest, appears antagonistic to the enculturation of institutional trust-based relationships.

5.1.4. Building Trust

Strategies for building trust have been developed and tested. In most cases, trust is built over time using ongoing actions and communications. Greenpeace, for example, commands high levels of public trust because of its openness, honesty, knowledge, expertise, and willingness to do unpopular things to retain its integrity, including admitting fault. If actors associated with Greenpeace make a factual mistake for example, they are generally quick to publicly admit the mistake and apologize (Bakir, 2006).

Third-party endorsement, from another already established trustworthy source, is an effective way to build trust. For example, celebrity spokespeople can be valuable third-party endorsers. Since expertise is bound in domains of knowledge or experience, celebrities can be both expert and inexpert, depending on the event or phenomena. Such endorsements are particularly effective and helpful in cases where a negative stereotype has been associated with an institution and public trust needs to be regained (Covello, Peters, Woitecki, & Hyde, 2001).

Defying a negative stereotype is critical to improving perceptions of trust and credibility in an organization. In the case of the Tylenol™ tampering case, Johnson and Johnson™ defied the corporate stereotype by responding aggressively to protect the public health and safety by removing all Tylenol™ products from the retail shelves. The company lost $100 million but gained trust that prevented further loss (Peters, Covello, & McCallum, 1997).

By contrast, Shell™ had worked to defy the stereotype of the oil industry by marketing itself as an environmentally friendly organization. However, its lack of wider consultation and transparency during the disposal of the Brent Spar oil structure reactivated stigmas attached to oil companies and the company lost much of its long established general trustworthiness (Löfstedt & Renn, 1997; Elkington & Trisoglio, 1996).

Similarly, in the last ten years the medical field has published research that urges full transparency and disclosure of medical errors (Kohn, Corrigan, & Donaldson, 2000). Medical doctors and researchers like Lucian Leape say patients need apologies before they can forgive and heal from their unexpected injury. Here, full transparency helped patients regain their trust (Schmidt, 2007).

In order to establish general trustworthiness through transparency (care, dedication, expertise, etc.) small group settings such as information exchanges and public workshops have been effective venues for communicating these characteristics (Covello et al., 2001). Also, active candor with the public can help maintain and expand trust with added public respect. For instance, Proctor & Gamble was damaged by incidents of sickness from tampons but actually gained public trust with its public- service advertisements warning about the dangers of Toxic Shock Syndrome (Bleeker, 1980).

Researchers have shown that positive policy information that sends a more direct message about the industry's overall intentions and general outlook is more effective than highlighting positive events in promoting trust in relation to negative events (White & Eiser, 2005). Thus, continually citing instances of general good performance to the public, including retrospectively naming specific examples of good performance after a negative event, may not be the best way of building general trustworthiness.

In light of epistemic trust's strong influence on the public's perception of risk, openly addressing scientific uncertainty should be a priority (see section 6). Classically, the precautionary principle stresses the importance of caution in cases of scientific uncertainty because perceptions of trust are decreased by actions or communications that reveal disagreements among experts. Even when the scientific community agrees about the low possibility of uncertainty, the effect on public attitudes may lead to lower rather than higher levels of trust (Wiedemann, Grutsch, & Schutz, 2006).

Scientists must also recognize and anticipate the media's insistence on reporting "both sides" of a story – even when there may not be (only) two justifiable "sides." For example, in the case of the measles, mumps, and rubella (MMR) vaccine reports, while the scientific community was in general agreement on the safety of the vaccines, the media's overall framework created the perception of a divided scientific community with two conflicting bodies of research (Speers & Lewis, 2004). As a result, the public began to lose trust.

Because science is never absolute, communicating openly and admitting uncertainty are important because people want to be informed and expect transparency (Short & Rosa, 2004; Frewer et al., 2002). At the same time, it is equally important to anticipate, acknowledge, and address "the other side" of an issue without necessarily legitimizing an opposing perspective. A trusted source that admits uncertainty may lose absolute authority but in turn enhances credibility. "The dilemmas facing risk communicators when they admit uncertainty cannot be eliminated. Yet neither can the need to acknowledge uncertainty" (Breakwell, 2000, p.117).

5.1.5. Whom the Public Trusts

Building trust involves knowing which people and organizations the public(s) trusts. Surveys indicate that citizen advisory groups, health professionals, safety professionals, scientists, and educators are perceived to have high-to-medium trust within areas relative to their fields (i.e., environmental, health, and safety [EHS] issues, etc.) and expertise. In general, consumer and environmental organizations were judged the most trusted sources, while government was considered the least trusted (Rosati & Saba, 2004). Other highly trusted sources include medical associations.

The name of an organization is also important for establishing trust. In a study by Hunt & Frewer (2001), a fictitiously named "Consumer Safety Council" gained one of the highest estimates of trust among respondents because its name conveyed trustworthiness. Neither Friends of the Earth nor Greenpeace rated nearly as high. Hunt and Frewer also found that food manufacturers, supermarket chains, and biotechnological companies all shared low estimates. To address lower trust levels, organizations should associate themselves with an organization of higher trust that holds the same position on similar issues in order to utilize trust transference, a phenomenon where sources takes on the trust and credibility of the highest trusted source with the same position on an issue (Covello et al., 2001).

Another way to build trust is through a highly-trusted individual. Because trust in individuals has a greater influence on perception than organizational trust, a person from a highly trusted organization may be judged more on personal actions and behavior than on his or her association with a trusted organization. Therefore, if a person performs admirably in public, that person may be regarded positively even if trust in his or her organization is lacking. This is especially important during a crisis. For instance, Bridgestone/Firestone Inc.'s choice of CEO, Jacques Nasser, as spokesperson during the tire recall crisis in 2000 was considered ineffective and distracting because of his accent, stiff manner, and name. Though he may have been capable of discussing the severity and complexity of the crisis, he did not convey credibility to the public and consequently did not build trust in the organization (Blaney, Benoit, & Brazeal, 2002).

The trustworthiness of various media and information distribution channels also has different public ratings. Just as government leaders were rated as having low trust and credibility ratings so do tabloid newspapers (Hunt, Frewer, & Shepherd, 1999; Frewer et al., 1996; Etzioni & Diprete, 1979). In Rosati and Saba's study, the most trusted media sources include TV documentaries, articles in high-quality newspapers, and consumer associations (2004).

With the new capabilities of digital media, governments have more opportunities for interacting more broadly with the public. Most governments now have web pages devoted to e-government efforts. However, users have not experienced the government as more inclusive or participatory through these sites. This may be due to Internet users who are more accustomed to creating as well as consuming information. It is possible these users may experience the government web pages as less interactive. Citizens expect e-government, like offline government, to use transparency and interactivity to build trust. Critics have argued that by using non-interactive web pages the government creates an impression of "building a screen bureaucracy that places citizens farther from rather than closer to bureaucrats" (Welch & Hinnant, 2003). This is a problem plaguing many institutions that define themselves using a passive web presence.

5.2. Fear-Based Arguments

Fear is a persuasive communicative strategy. We live in a fear culture and the purveyors of fear include members of government, industry, non-governmental organizations and the media. In short, fear motivates and changes attitudes and behaviors. Fear enters into persuasive appeals as a consequence of many factors. For example, communication experts use fear appeals in crisis communication when the goal is community protection at all costs.

However, fear is a fickle seducer. Fear appeals cannot be sustained unless the magnitude of the fear is increased and reinforced. Even in these circumstances, a saturation point is reached where the fear appeal becomes ineffective and counterproductive. Those individuals subjected to pervasive and redundant fear appeals begin habituating and ignoring the message. In some instances, they begin identifying its manipulative function and the message has a boomerang-like effect (Rogers and Mewborn, 1976).

The traffickers of fear are often convinced that using fear tactics to convey messages is needed in order to compete for attention and interest in a cacophonous world. For example, the television media in the 21st century must compete with not only with cable offerings on hundreds of channels but also with a nearly endless list of options found in the digital media of the Internet. Given the multiplicity of venues vying for attention, fear is often regarded as an effective means to gain advantage. For example, teasers are a common occurrence in episodic serial television and news programming.

5.2.1. Dread

Some risks are simply dreadful and terrifying by nature (Covello & Allen, 1988). The classic example of dread is nuclear war. This apprehensiveness is culturally enhanced by textual and visual evidence from weapon tests in Nevada and the blast- lands of Nagasaki and Hiroshima. This sense of undifferentiated dread has transferred to almost all things nuclear, especially nuclear power plants and the waste they produce, and has been amplified since Chernobyl and Three Mile Island to include all forms of nuclear power.

An example of this comes from nuclear magnetic resonance, best known to the general public as magnetic resonance imaging for medical diagnosis. The word nuclear was eventually dropped from this technology's name to insulate this medical technology from the dread transference associated with the word nuclear (Cochran, 2004).

A similar phenomenon may be occurring involving the use of the prefix "nano" as a marketing device for products involving some aspect of nanotechnology, such as cosmetics and sunscreens, or comparative smallness in size, such as the iPod Nano.

5.2.2. From Social Cascades to Moral Panics

Fear can be contagious. This contagion is passed on through social cascades, where people pay attention to the fear expressed by others in ways that can lead to rapid transmissions of beliefs, even when false, that a serious risk exists. This social belief heuristic whereby fear triggers a chain reaction involving one of more agent(s), such as non-governmental organizations and the traditional media, leading to moral panic. Cohen (1972) has referred to moral panics when he explained how a minority group might be perceived as a “menace” to society because of exaggerated misperceptions. This concept also explains how non-rational behavior can function as a consequence of fear mongering (Sunstein, 2005). While long-term fear-driven social cascades may prove ineffective or even counterproductive, their short-term impacts should not be ignored because if not addressed, they can lead to acute and unnecessary non-rational moral panics.

5.2.3. Verify and Respond

Responding to fear appeals is straightforward. Fear claims must be scrutinized and responses must be quick and direct. Much like the response of a political candidate to negative campaigning, a quick and direct rebuttal to a fear appeal can be an effective response strategy. Although responding to a fear appeal may lend some limited credibility to the claim, it is vital that appropriate, strategic, audience-specific responses be employed to ensure other perspectives are available for the audience to consider - failing to respond to a fear appeal is worse than rebuttal. Communicators should be mindful that impulsive responses that do not adequately address audience, situation, or other qualitative, formative issues can also lead to cascades and moral panics.

5.3. Belief-Based Arguments

Belief-based arguments are especially troublesome because beliefs are used as warrants for claims and counter-claims. In this situation, with both sides having affirming messages (e.g., both pro-choice and pro-life), any accommodations and compromises are seen as sinful responses by those promoting such belief systems. For example, belief-based arguments in debates over abortion, embryonic stem cell research, and gay marriage currently exist, with speculations that nanotechnology could be added to the list in the near future (Scheufele & Brossard, 2008).

5.3.1. Nanobiotechnology and Synthetic Biology: Horizon Issues

Nanobiotechnology and synthetic biology may already challenge existing belief structures in three ways:

- Researchers are perceived as playing God.
- Research is perceived as engaging in and promoting unnaturalness.
- Research is perceived as promoting and creating the risk of dehumanization.

These belief concepts factor into the principles upon which our attitudes and values are built. For nanobiotechnology, other unanticipated stakeholders may be active including religious spokespersons and associated advocacy groups. When discussing nanobiotechnology as compared to other areas of nanotechnology, conversations based upon religious beliefs may also increase in emotional intensity and volume. In these types of debates:

- Scientists may be demonized as philistines. Inevitably, the image of Frankenstein will be raised and used against science;
- Data may be limited on understanding how resilient religious opposition can or will be in these discussions;
- Some scientists and advocacy groups will have anticipated potential opposition, levels of vehemence, degrees of persistence, or alternative narratives for nanobiotechnology while others will have not, but all will be participating in the discourse regardless of their preparation.
- More social science research will be needed to explain and predict the capacities advocacy groups and non-government organizations have to coalesce into full social movements that influence consumer behavior and public voting.

Flexible and responsive rhetorical tactics maximizing accurate information while debunking hyperbole and falsehoods should be fully tested in each medium used. For example, recall that traditional media (newspapers, TV) are no longer the news media of choice for a growing number of Americans (see section 3). However, simply relying on digital media without sufficient research and data on public engagement in this medium could be insufficient and shortsighted as well.

5.3.2. Visions of the Nanobiotechnology Future

Significant advances in biotechnology sustained and accelerated by parallel developments in nanoscience suggest powerful visions of the future. Human concerns about disease and health suggest positive interest in and even reception of applications in nanomedicine from nanobiotechnology. However, public responses to cosmetic enhancements, genetic manipulations, and life-prolongation driven by nanobiotechnology may vary depending on the application. These developments tap into religious values and what defines us as human. The characteristics that make us human determine and are determined by religious belief systems, national identities, peer groups, occupations, and individual character and self-identities.

Generally, a change in a fundamental belief is due to an epiphany precipitated by a major or critical life event. Whether nanobiotechnology will have a watershed event that triggers a need to engage in public discourse remains speculative at this point. Some issues will be transferred into and emerge from the world of fiction, rumor, and speculation. The challenge will involve using arguments and warrants for claims built on belief systems, beliefs systems that rarely change, and when they do, not easily.

5.4. Summary and Conclusion

Within the current culture of fear coupled with the public trusting institutions less and less, specific strategies should be employed to ensure that members of the public are able to be informed, understand, and appropriately weigh information in order to respond to risk messages.

Epistemic trust is more important to public perception than social trust. Meeting epistemic concerns requires a demonstration of valid scientific theories and research data—a way to inform members of the public about nanotechnology so that the potential and real risks and benefits may be weighed thoughtfully.

When communicating to the public about fear claims that have already occurred, it will be important to scrutinize and vet responses quickly and directly while being mindful of the delicate balance of responding to the fear itself possibly lending credibility to the claim. Nevertheless, risk communication must be appropriate and audience-specific so that fear is not increased and specific groups are not offended.

Finally, the advances of nanobiotechnology will attract belief-based claims. In response, dedicated social and behavioral science research and response strategies will be needed.

SECTION 6. PUBLIC ENGAGEMENT

An important issue in risk communication is *who* decides *what.* While management and risk assessment experts can determine how risky a work environment may be, input on how much risk is acceptable should also come from those who are exposed to the risks. Exposure assessment is complicated by informed choice factors such as cognitive ability and income level.

Public participation has become an adjunct to regulation and industry. The desire to include the public in participatory activities largely stems from concerns that traditional policy-creating processes can be enriched through increased stakeholder participation (Hamlett, 2003; deLeon 1990, 1997; Mayer 1997). Furthermore, public participation is used for input into research and policy development processes. However, disagreements exist on how to incorporate public input, especially when the public has very little knowledge and understanding about nanoscience (see Appendix B).

Some goals of public participation are:

- to find ways for the public to participate within a democratic process,
- to inform the public, and
- to enable meaningful input in the processes of science policy.

As nanoscience continues to grow as an interdisciplinary field, public engagement may provide the opportunity for heightening public awareness and understanding of scientific issues and for the development of greater participatory democracy governing science and technology policy making.

6.1. Stakeholders

Stakeholders are people or organizations who have vested interests in the actions, applications, and impacts of policy makers and researchers. They try to influence research agendas, design and produce consumer products, and write policy. For applied nanoscience, stakeholders are a heterogeneous group, with scientists, technologists and industry leaders at the forefront directly developing the science and creating technological uses for scientific discovery. Of course, members of the general public are also stakeholders, especially when specific events bring them and their interests within the scope of nanotechnology projects. At the same time, there are many others whose careers are (or soon will be) directly linked to nanoscience, including company managers, bureaucrats, and insurance agencies with whom production and financial risks associated with nanoscience are tied. Public service groups, media representatives, academic commentators, manufacturers, and consumers within the public at-large are stakeholders most closely tied to the implications and proliferation of nanoscience products (Berube, 2008b).

The concept of "the stakeholder" transverses through media, government agencies, and public advocacy groups' publications. In one sense, all entities with a vested interest in the field of nanoscience (be it financial, social, ethical, etc.) are stakeholders. What this means is that stakeholders include both those dedicated to the expansion of applied nanoscience and those with opposing views (Berube, 2007; Bonnafous-Boucher, 2005). As nanoscience has progressed, there has been an increased desire by policymakers to engage stakeholders positively and in accordance with the 2007 National Nanotechnology Initiative Strategic Plan (Berube, 2007; National Science and Technology Council 2007).

6.2. Classifying Engagement Models

Models can be classified by their goals. Deliberative polling, consensus conferences, citizen technology forums and citizen juries seek to democratize the processes of agenda setting of scientific research and development as well as the development of science policy. Smaller models include science cafes and citizen schools whose focus is concerned with heightening the awareness and understanding of members of the public concerning specific science issues.

6.2.1. Deliberative Polling

Deliberative polling bridges the gap between uninformed public opinion and informed public perspectives on a topic (Sturgis, Roberts, & Allum, 2005; Luskin, Fishkin, & Jowell, 2002; Guston, 1999; Fishkin, 1991, 1995, 1996; 1997). Fishkin (1996) argues that this model demonstrates what general public attitudes and opinions would be if the public had the opportunity and motivation to be engaged stakeholders. The goal is to disperse findings among the population at large and raise public awareness and understanding of an issue. This model measures attitudinal and informational changes in a quasi-experimental setting by soliciting and incentivizing a random sample of citizens to read, listen, and/or debate some issue. Generally, presentations involve balancing competing perspectives from experts and political figures over the course of several days. In some instances, sessions are videotaped and organized into a program for public access via national television broadcasts or Internet webcasting.

Critics of deliberative polling argue that the financial, administrative, and infrastructure expenses associated with hosting national conferences and crafting resulting videos do not justify their limited impact on the public at large. Additionally, there is limited evidence available that shows how these activities generate long-lasting changes in opinions and attitudes of participants or significant or sustained impact on policy makers. At best, this model demonstrates a quantitative understanding of the public's attitudinal change using simple pre- and post- tests after completing a short primer about an unfamiliar topic.

6.2.2. Consensus Conferences

The "consensus conference" follows the 1987 Danish Board of Technology's efforts whereby public stakeholders are invited to participate in panel discussions over three weekends roughly one month apart. Like many of the current public engagement models, the goal was:

- to increase public awareness and understanding of scientific issues; and
- to produce a democratic mode of operation for the creation of research and development agendas and policy-making.

Typical uses of this model include 12-15 recruited members of the public who meet on two or more separate occasions to deliberate about a topic. In a three session model, prior to the first two sessions, the panel members are provided with written background information concerning the topic matter and are encouraged to discuss and question the information in face-to-face interactions. During the first meeting, a group discussion about the topic is led by a trained facilitator. The second session involves the public participants asking questions of a panel of experts holding a variety of viewpoints The third and final session culminates with the citizens collaboratively detailing a report of their findings and recommendations, which is then disseminated to the public, governmental agencies, and policy makers (Powell & Kleinman, 2008; Hamlett & Cobb, 2006).

Despite their limited influence on policymakers and agenda setters (Powell & Kleinman, 2008), consensus conferences provide opportunities for motivated individuals and collective groups of citizen panels to increase their knowledge of a particular topic and heighten civic engagement through increases in knowledge. In addition, they may function as trust-building exercises.

Similar to deliberative polling, the financial and organizational costs versus the usefulness and limited demonstrable impacts on the general public may make the requisite financial support needed to provide the conferences difficult to justify. Additionally, it may be unrealistic to presume that a group of forums on the same issue would produce significant impacts on the research and policies surrounding the issue. While this model affords scientists, technologists, and policy makers interactions with members of the public, this model does little to ensure how the recommendations from the empanelled citizens will be used.

6.2.3. Citizen Technology Forums

This modified consensus conference model draws from the National Forum on Science and Technology Goals recommended by the Carnegie Commission on Science, Technology, and Government (Guston, Woodhouse, & Sarewicz, 2001). The current model designed by Patrick Hamlett of NCSU uses a small panel of citizen jurors who meet both face-to-face and via the Internet (keyboard-to-keyboard). The groups are situated in different cities in order to provide geographical diversity of opinions. Members deliberate with experts while being led by a trained facilitator (Hamlet & Cobb, 2006). Similar to the typical consensus conference model, the final meeting culminates in the crafting of a paper that assesses the topic and situation at hand.

The substantive difference between "citizen's technology forums" and consensus conferences is the use of Internet technology as a means for communication without the necessity for in-person face-to-face communication. While this may help to deflate some of the costs (travel, compensation, administrative fees, etc), this model may not prove as valuable a tool as face-to-face interactions for heightening the public understanding of science, for efficacy in engagement, and for instilling trust in the scientific community. Similar to the consensus conference, the outcome of this model, an assessment paper, does little to ensure that the recommendations from public members will be taken into consideration post-forum.

6.2.4. Citizen Juries

Another variation of consensus conferences is "citizen juries." Conducted since 1974 (Jefferson Center, 2008), citizen juries involve a "representative" randomly selected panel of 18 or so citizens/jurors who meet for four or five days. Similar to a courtroom, the panel hears testimony and arguments from expert witnesses in moderated sessions and deliberates in order to provide recommendations at the conclusion of the event (Berube, 2006). An advisory committee assists the jury in order to help avoid bias.

In 2005, an independent citizen jury met over five months in the United Kingdom to deliberate about the impacts of nanotechnology (NanoJury UK, 2005). The process began with first-round discussions over a local crime issue and moved in its second round to nanotechnologies. They delivered 20 recommendations (only four of which were judged significant) about how nanotechnology research and development should be approached in the future. Unfortunately, the jury lost many participants going into the second round.

These forms of engagement are designed to provide a more representative sample of the public as well as a greater understanding of the topic. In both instances, the relevance of the exercises was dependent of whether regulators interfaced with the participants and considered their conclusions. The effectiveness of these projects remains generally undemonstrated.

6.2.5. Science Cafes

Science cafes are scientific talks with question and answer sessions hosted in local coffee shops, restaurants, bars, or other public locations. Here, an expert in a particular field leads a discussion concerning an issue and encourages the public to enter in debate and consensus building. These gatherings rarely culminate in any formal reporting; rather they serve as an informal education outreach activity for voluntary self-selected ad hoc participants who have the time, means, and motivation to participate (Berube, 2006).

Unlike many other models of public engagement, this model primarily focuses on disseminating information rather than creating a democratic process of decision-making for research agendas and policy. In this model, difficulties exist in assessing the quality and balance of the information provided.

6.2.6. Citizen School

Pioneered by Christopher Toumey, the "citizen school" functions as a localized example of continuing education outside of the traditional institutional framework. The goal is to educate a small portion of the population about science and technology. In citizen schools, a self-selected sample of the population engages in classroom discussions with a teacher who is typically a scientist in the field.

While valuable and informative for participants attending these schools, participants are typically not representative of the general public. Additionally, there is no verifiable mechanism available to validate or assess the quality of information or biases that may be presented within the classroom setting by instructor or held by participants (Berube, 2007).

6.2.7. Assessing the Models

Public engagement activities reflect a genuine desire to incorporate public opinions into the decision-making processes for emerging technologies. Engagement models largely seek to raise society's capacity for:

- understanding science issues;
- collaborating with the scientific community; and,
- incorporating public opinion into the development of research agendas and policy-making.

Some public engagement models are used to test rather than produce meaningful information. Others use public engagement as a tool to quench the public's curiosity or defuse suspicions. And others use public engagement as educational outreach to inform those who have the power to make policy and affect the daily lives of the public. The difficulty lies in matching the abilities of these models to serve their desired functions. While well-intended, often these models are deficient in reaching their idealized goals because of inadequate sampling procedures, lack of demonstrable impacts, and cost.

It is important to gain access to representative groups from the public when trying to make claims about the attitudes and opinions of the general population. Unfortunately, current public engagement and deliberation models use sampling techniques that are not sufficiently representative of general populations. For example, participants who are self-selected individuals who have the time, the means, and the motivation to partake in lengthy conferences and workshops are far less representative than members of the general public. The consequence of models that use self-selected samples from special populations is that the conclusions drawn do not represent the attitudes and opinions of the general public. Furthermore, if the intent of compensating the public to participate is to reduce financial or inconvenience cost of participation, this incentive may alter the motivations of the participants to participate and have little effect on reducing sampling error.

In order for a public engagement to be effective, participants must engage the broader general public. Additionally, there is no evidence that individuals who participate in these models become group opinion leaders, heighten the awareness and understanding of an issue for the broader public, or influence policy. **Patrick Gehrke**[6] also raises questions about the democratic nature of current public engagement models and their underlying assumption of democracy existing everywhere individuals are deliberating. These models in practice create artificial democracy through artificial contexts and artificial deliberations with little "ecological validity" and low inferential reliability to other reality-based contexts. For more realistic engagement activities, we need to study the public at existing deliberation sites where they traditionally assemble, such as church halls and club meetings.

Current public engagement models create deliberative communities that are not representative of organizations and groups within the population at large (see section 4.3.3 concerning the "Big Sort"). Some current engagement practices can be beneficial for participants (both experts and non-experts) but their use as tools for generating data to inform research agendas, science policy, as well as creating advanced rational discourse and debate on science and technology remains uncertain. For true democratic deliberation, the discord between how engaged stakeholders deliberate in their known communities within society and how strangers in idealized settings deliberate with trained facilitators and experts providing structured information will need to be settled.

Coordinated efforts will need to be organized to develop and assess engagement techniques and outputs if broader spectra of people are to participate in the world of new technologies. Additionally, engagement models will need to inform a greater portion of the public and help a diverse group of individuals make rational judgments about science and emerging technologies. Methods including ethnographic methods in existing communities already involved in natural deliberative processes will be useful in gathering and understanding realistic data about stakeholder deliberation (see section 3.4 concerning digital media research initiatives).

6.3. Public – Private interface

A term "upstream engagement" refers to involving the public in the early processes of product design or regulatory policy making rather than post-facto participation. To this point, **John Stone**[7] has developed an approach where upstream engagement involves transferring local knowledge to those situated best to use it in production and regulation.

6.3.1. Defining Streams

For industrial production and regulation, linear stream analysis involves public engagement along a continuum of product development to eventual consumer acceptance (Fisher et al., 2006). Policy analysts describe this inception of technology theory through:

- gaining funding,
- setting a research agenda,
- completing research and development,
- applying research findings, and
- marketing the technology.

6.3.2. Downstream and Upstream

Conventionally, the public is engaged "downstream" from setting research agendas and making policy and constrained to inquiring about the risks and impacts of the technology post-production (MacNaughten, Kearnes, & Wynne, 2005). At this downstream position, the public is engaged to assess consequences and serves more as a consumer having little or no input.

Upstream public engagement has been conceived as incorporating citizen input closer to the setting of research agendas, a task typically reserved for scientists and, in venture capital situations, shareholders or investors. In this scenario, an assumption exists that these citizens will have access and interface with regulators similar to the Denmark consensus conferences.

Technology has historically been framed as a "black-box." Rejeski (2005) has described an ongoing and increasing public suspicion of new science, writing that we need to be cautious about "...a public that has grown more suspicious of both public and private sector motivations concerning technological advances and a scientific community that remains largely isolated, and often oblivious, to public concerns" (p. 5). In response to mediating increasing public suspicions, some have advocated upstream public engagement approaches to create a democratic approach to the production, commoditization, and regulation of science and technology (Wilsdon and Willis, 2004).

This approach has one axiom: the public should have an opportunity to influence policy makers, funding agencies, researchers, and scientists from the inception of emerging and converging technologies such as applied nanoscience. Therefore, the primary goal of public engagement should be incorporating the public voice into the planning process before action plans are decided by scientists, industry, and policy makers. In general, these approaches advocate for a public understanding of societal concerns prior to the "risk and impact" phase of downstream social inquiry, ideally alleviating some of the panic and stigmas associated with public perceptions of applied nanoscience.

Upstream engagement has been theorized as a way to increase trust among governmental agencies, non-governmental organizations, and the public (in various representations) through strategic communication and properly executed engagement activities. Upstream engagements may uncover possible solutions to unforeseen problems due to public scrutiny and produce a more keen anticipation of societal issues from instantiation (MacNaughten et al., 2005). However, what remains difficult is how to provide significant upstream influence when few non-specialists have the expertise, insight, cross-disciplinary experience, or policy access to engage technological development at upstream points.

To these points, Kuzma, Romanchek, & Kokotovich (2008) have proposed an "upstream oversight assessment" (UOA). This approach identifies data, risk assessment, regulatory policy, and engagement needs to help society and policymakers anticipate and prepare for challenges posed by applications of agrifood nanotechnology using applied nanotechnology case studies as a method to reveal application areas needing further study and attention.

6.3.3. Midstream Modulation

Fisher, Mahajan, & Mitcham (2006) have also suggested including the public during research and development processes that come in-between the creative inception of technologies and the full creation of products for public consumption. This variation of the upstream framework emerged in response to critiques of upstream models that were implemented too early in the process and downstream models that are implemented too late in the process to be effective.

Applied to the context of the midstream, this concept of modulation can help guide internal attempts to conduct and implement R&D with an eye toward subtly and creatively shifting ongoing, nested interaction among techno-scientific actors and networks (p. 492).

Fisher et al.'s midstream model draws heavily from "constructive technology assessment" (see below). Midstream modulation already occurs in R&D processes although participation is generally limited to research and corporate stakeholders. Including stakeholders from outside the traditional R&D organization could purportedly increase public influence over research trajectories and allow more timely responses to technologies that may not claim to influence science or technology at the inception stage.

Unfortunately, similar uncertainties associated with any early stage assessment occur in this model. While some benefits to integrating public stakeholders into academic research projects may exist, there is less evidence that these approaches effectively translate in business settings. For example, the process may be most effective in product lines, such as nanomedicine, which may be highly affected by consumer demand. Attentiveness to relevant variables by naïve actors is also a challenge. While dialogue with public stakeholders might be desirable, the output from midstream modulation involving the public may also be limited by precompetitive confidentiality issues and concerns associated with slowing R&D processes for those who choose to incorporate midstream modulation participation. How many public stakeholders to involve and where they are situated, will also need to be addressed.

6.4. Assessment Approaches

Technology assessment is especially challenging. The U.S. government's official Office of Technology Assessment was retired in 1995. While some claimed the office as ineffective and expensive (Lambro, 1980), others lamented its demise (Bimber, 1996). To meet the needs of technology assessments, some have come forward with creative means to assess ongoing technological developments such as nanotechnology.

6.4.1. Constructive Technology Assessment (CTA)

"Constructive technology assessment" (CTA) is premised on the need to transition from the downstream model of technology engagement to public engagement prior to the solidification of research agendas (see Rip, Misa, & Schot, 1995). This model uses early and controlled experimental methods to identify possible unforeseen downstream complications (Rodemeyer, Sarewicz, & Wilsdon, 2005; Guston & Sarewicz, 2002).

6.4.2. Real-Time Technology Assessment (RTTA)

The goal of "real-time technology assessment" (RTTA), an extension of a CTA, is to engage the public to alter the current structure of industry R&D. RTTA advocates depict the R&D industry as currently isolated from the public and try to influence the process of knowledge creation itself (Guston & Sarewicz, 2002). By examining the interface of science, policy, and public input, they believe adjustments managed through incremental stages can be made in the R&D process.

This model collects data in multiple forms including:

- reflexive tools (opinion polling, focus groups, and scenario development) to gain insight into stakeholder values and diverse potential outcomes,
- content analysis and survey research to determine how knowledge and values evolve over time, and
- socio-technical mapping in order to gain retrospective and prospective accounts of interaction between technological advances and the public at large (Guston & Sarewicz, 2002).

Idealistically, RTTA is a form of "purposeful trial-and-error," an iterative process that is continuously doing retrospective analyses in order to manage the unpredictability of technological development in a system. However, Cash et al. (2003) suggest RTTA has significant hurtles. First, building of a data set may be difficult, especially given their observations that traditionally "perspectives of a decade or more may be necessary to reliably evaluate the impact of science, technology, and ideas on issue evolution" (p. 1). Second, there are the challenges imposed by communication norms and protocols between disciplines including different sciences, like biology, physics, chemistry, and policy studies. Although Jasanoff (1990) has claimed that blurred interdisciplinary boundaries lead to better policy, others have argued that this blurring may have facilitated the recent trends of the politicalization of science (Mooney, 2005). Third, boundary theorists have their own language (boundary objects, standardized packages, and boundary organization) that may limit accessibility. Additionally, they tout the utility of an idealized boundary organization being able to balance strategies using science and political principals as a resource to humanize and socialize science and technology without acknowledging the costs involved in such research (Scott, 1992).

Communication between fields is dependent on many field-specific characteristics including the motivation of the actors in those fields at their intersecting points. Similarly, translation between and among fields is fraught with errors and densely layered with field-specific experiences and presumptions, which can be idiomatic and at times untranslatable. While mediation activities involving experts and decision makers may be needed to navigate contested areas, it is unclear how realistic these expectations and outcomes are given the different roles these actors play in different science policy discussions.

6.4.3. Assessing the Assessment

How will we know if public participation activities are effective? How public is public participation? How upstream is upstream engagement? How real-time is real- time technology assessment? The work of public engagement is besieged with new terminology with self-defining and often favorable descriptors that may mislead those less familiar with the field. For example, "life-cycle analysis" is a methodology while "functional life-cycle analysis" is an ideological term misnomered as a methodology.

6.5. Summary and Conclusion

Democratic science does not demand science by referendum but instead "creat[es] institutions and practices that fully incorporate principles of accessibility, transparency, and accountability. Science [must] be popular, relevant and participatory" (Guston, 2004, pp. 25-26). While current approaches for enabling a more democratic science may need further review, some call for a new Office of Technology Assessment to restore "the policy-analytic balance between Congress and the Executive Branch in matters science and technological" (p. 27). The call for "increased opportunities for analysis, assessment, and advice-giving through the use of deliberative polling, citizens' panels, and other participatory mechanisms" (p. 28), although expensive, may improve the process of decision-making, but not necessarily produce better policy. Some suggestions include involving members of the public in grant review panels, public outreach expectations in research grants, and enhanced transparency and accountability of expert deliberation.

Nevertheless, caution must be exercised. First, involving more members of the public in science and technology decision-making assumes their interest and commitment. Second, selecting who represents the public is challenging. Hiring professional "publicans" to serve as public representatives may raise concerns about representativeness and their role in democratic institutions. And third, the field of public engagement has boundary features worth examining. What remains unclear is how effective all of these efforts are in advancing responsible risk communication. This does not mean they cannot though they cannot serve as trust building exercises.

Recent efforts to narrow focus to "upstream" engagement, although well intended, may be too simplistic (Wilsdon and Willis, 2004). Upstream and downstream engagements are inextricably linked to one another in any successful engagement effort. Upstream engagement with boutique populations can be an invaluable tool for identifying key issues and concerns for downstream communications efforts. As researchers have shown, most citizens do not possess the motivation or expertise to participate in any upstream engagement efforts (McLeod et al., 1999). As a result, broad impacts on public attitudes toward and understanding of new technologies are a function of when and how the issue is presented and plays out in mainstream media (Scheufele, 2006). With GMOs as an example, upstream public engagement efforts could have been used to potentially avoid commercial pitfalls for early consumer end products such as Flavr Savr, where downstream opinions among the general public were mostly a function of well-researched and well-implemented strategic communication efforts such as Frankenfood frames and other communication devices.

SECTION 7. NEW RISK FINDINGS

The field of risk communication is constantly in flux as new situations arise that need to be disposed. Disagreements over risk are common as science progresses into new areas that have the potential to affect public life. Risk communication researchers may likely face "hot button issues" that have the potential of changing into bottlenecks, road blocks, and/or trigger a wide range of reactions. Risk communicators should be mindful of issues as they prepare to communicate with the public about risk.

Whoever controls the communication and defines the risks controls the solution to the problem at hand. "Defining risk is thus an exercise in power" (Slovic, 1999, p. 4). Communicating risk is a tool for empowerment. Who explains and thus defines the costs and benefits associated with new technologies such as nanotechnology? The experts? With the public? With policy makers? Or as a cooperative exercise involving all the stakeholders?

7.1. General Approaches

The following are broad theoretical frameworks of risk communication on science and technology that are supported by quantitative and qualitative analyses and should be informative for risk communication of nanotechnology.

7.1.1. Scientiating the Public: The Deficit Model

Science has been typically portrayed, and perceived, as an objective process that uncovers natural truth through strict empirical methods. The nature of science is discovering truth by decreasing uncertainty. In areas where high levels of uncertainty remain, scientists use intuition and hypotheses derived from their best understanding of the subject matter. It is the collective pursuit of uncertainty reduction that enhances the validity of scientific discovery.

What remains difficult for the general public to comprehend is how uncertainty can exist after the laboratory experiment is completed *and* how disagreements between scientists about interpretations of research data can occur and be sustained. While scientists regard debate as healthy discussions that forward science, the public tends to negatively interpret these debates and peer-to-peer challenges as unproductive, contentious, and unsuccessful.

Furthermore, communicating risk becomes difficult when the public's interest and support for science wanes when compared to other more personally relevant events like the price of gasoline and home heating oil. This will be increasingly problematic if declining interest parallels the rapid emergence of new science and technology. This lack of interest in and support for science can instigate a hostile public that gathers information from a sometimes equally hostile media. These negative reactions from the public can also encourage experts to remove themselves from the social aspects of the discourse (Freudenberg & Pastor, 1992).

To improve communication between expert and public spheres, some advocate improving the science literacy of the public. Deficit model proponents suggest: if individuals have a comprehensive understanding of relevant science information, they will cease to fall back on irrational fears or embedded beliefs about science. The science deficit model posits a positive relationship between science knowledge and public attitudes toward science (Allum, Sturgis, Tabourazi, & Brunton-Smith, 2008).

There are numerous challenges to this model. First, an implicit assumption exists that non-rational public fears are based on a lack of scientific understanding. Researchers have shown that individuals' scientific knowledge is not as important a predictor of risk perception as qualitative social factors such as social trust or worldview (Slovic & Peters, 1998). Thus, socio-cultural associations may carry more influence about what individuals select as the norms for their risk estimations.

Second, it is extremely difficult to establish a threshold for the amount or type of information that is required for public science literacy. For example, the pertinent knowledge required to adjust attitudes around a science controversy can be highly contextualized (Hayes & Tariq, 2000). Additionally, threshold values of science knowledge have not been determined to see if knowledge proceeds "step-wise" or as a threshold where information and its density must reach a certain level before affecting attitudes.

Third, there is no guarantee that understanding science leads to better impact assessments or decision-making. Opponents have argued that high degrees of scientific literacy seem to inoculate scientists from recognizing or properly understanding the full influence or effect of new technologies. They suggest that those who best represent public interests and who recognize broad impacts of technologies are often those with sufficient but not entirely fluent (specialist) scientific knowledge (for an elaboration of this point, see Kinsella, 2004).

Defining (and measuring) knowledge and attitudes and determining the relationship between the two is not straight forward, In a meta analysis, Allum et al. (2008) used a multi-level modeling scheme and identified a small and weak positive correlation of knowledge to attitudes.

Science experts often support this deficit model and feel that increasing scientific knowledge—producing more science facts—will increase acceptance of science and technology. Although this strategy may intuitively make sense, as research and tracking of such efforts has shown, and while there are many good reasons for increasing science literacy, enhancing public participation in science is not one of them (Wynne, 1991).

How can we communicate about science to an uninterested public? Is providing the general public with more information effective? Does content knowledge in itself provoke attitude change? Wynne (1992) conceptualizes civic scientific literacy to include the following:

- formal contents of science knowledge;
- methods and processes of scientific inquiry; and
- ways that sciences are institutionally embedded, patronized, organized, and controlled.

In general, technical knowledge may be necessary but not sufficient to promote greater appreciation for science and technology. Public participants do not need to know everything in order to form attitudes favorable or otherwise toward science. Additionally, while identifying the critical amount of knowledge may be challenging, the amount is likely much less than what might be presumed.

The efficacy of risk communication strategies is not necessarily improved through an increased volume of content information. Distributed science information is likely to be most effective when content information is delivered within the context of specific examples, science methodologies, and science organizational and power structures.

Efforts toward increasing science literacy (e.g., methods, institutions) for nanotechnology may develop greater consumer trust and better facilitate the intersection of communication between the public, science experts, and science- oriented audiences. However, the challenges of risk communication about science and technology are complex and layered.

7.1.2. Boomerang and Backlash Effects

A methodical approach will need to be taken when addressing the counter-intuitive conclusion that more information may increase negative perceptions about nanotechnology, regardless of the sources or the perspectives of this information. In some ways, this counter-intuition supports socio-linguistic research showing that when people are repeatedly exposed to information, they intuit a suspicious implication about the intentions behind the multiple and repeated messages. In other words, people will suspiciously intuit the over-emphasis of the safety of nanotechnology as uncertainty about the safety.

A second explanation is confirmation bias (see section 4). The Yale Cultural Cognition Project team noted that

> ...it does not appear that learning more about nanotechnology tends in general to make people more favorably disposed to it. There is a positive relationship between how much people know about nanotechnology and the belief that its benefits outweigh the risks. But when people who know little or nothing are supplied with more information, they do not become uniformly more favorable; some form a more positive impression, some a more negative, depending (again) on their values (Kahan, Slovic, Braman, Gastil, & Cohen, 2007, p. 2).

Using a cultural model, not unlike the early work of Douglas and Wildavsky (1982), Kahan et al. (2007) validated the confirmation bias whereby individuals were found to believe arguments that simply confirmed a prior-held belief: "People who hold largely individualistic and hierarchical cultural outlooks tend to see nanotechnology as more beneficial. People who hold largely communitarian and egalitarian outlooks, in contrast, tend to seen nanotechnology as more risk when exposed to that same information" (Kahan et al, 2007, p. 1). It should be noted that other bipolar semantic differentials could be substituted, such as conservative/liberal and Republican/Democrat.

As an example of bias, a series of **National Citizens' Technology Forums**[8] reported a negative relationship when engaging individuals from the public in discussions about nanotechnology for human enhancement. Perceptions about human enhancement involving nanotechnology grew more negative after a public engagement exercise. However, because of methodology limitations, this study's conclusions should be cautiously interpreted. For example, the publics' 82 percent (82%) pre-test favorability ranking toward nanotechnology was exceedingly high due to the fear of death (the most powerfully recorded bias). The post-engagement rankings of 66 percent (66%) actually demonstrate the capability to compensate for this bias through engagement. Thus, the difference of 16 percent (16%) may have been due more to the result of inflating effects of the bias than to the information provided on nanotechnology per se. But for the bias, the pre-test rankings would have been much closer to the post-engagement figure. Here the public engagement exercise dampened bias through discussion and showed that providing more information may not be as important in assessing public perceptions about nanotechnology as an affect heuristic (Kahan et al , 2007).

7.1.3. Uncertainty Effects: Fuzzy Risks

Risk is built on uncertainty. If a hazard were known, there would not be a true risk, there would be certainty. Researchers and lived experiences demonstrate that uncertainty and risk ambiguity remain important, difficult, and at times a contradictory phenomenon in public communication such as in the areas of science and technology (Kahneman, Slovic & Tversky, 1982). Moreover, uncertainty is a major obstacle to decision-making (Gigenrenzer, 2002; Orasanu & Connolly, 1993; McCasky, 1982; Brunsson, 1985; Corbin, 1980; Thompson, 1967).

Instead of defaulting to probability estimates when determining hazards and risks, fuzzy set theory uses membership functions and linguistic parameters to express vagueness in issues *and* probability theory, a stochastic approach using probability functions to describe random variability in parameters, especially environmental ones (Darbra & Eljarat 2008). These systems make valuable contributions to uncertainty analysis. Darbra and Eljarat argue "[t]he flexibility of fuzzy logic to express results in a natural language, in line with human reasoning, together with the possibility of dealing with uncertainties makes it highly recommended as a tool for use in communicating risk" (p. 384). This model is highly affected by subjective variables such as human judgment. Thus while probability remains relevant probability alone is insufficient when modeling complex health and environmental parameters, especially when boundaries are not defined clearly. These observations lead Darbra and Eljarat to conclude that "[a] combined approach may therefore be the best solution to deal with the uncertainties. Treating each parameter according to its nature and [available] information could help to obtain a more realistic result for risk estimation" (p. 384). Huang and Ruan (2008) have drawn similar conclusions based on their work in fuzzy risk estimation involving probability of a fuzzy event, expert evaluation, and information diffusion (p. 684), where they have developed an average weighted algorithm to update fuzzy risks in terms of fuzzy probability distributions (p. 692).

There is a tendency to default to probability estimates when determining hazards and calculating risk. However, a threshold exists where the degree of uncertainty is so great that probability estimates are nearly indefensible. Fuzzy logic and risk estimation may be more than mots du jour. They may offer useful guidance to both estimating risks and designing communication algorithms and models. For example, what happens to a risk analysis when data are simply unavailable due to randomness or incompleteness? In these cases, practical, comprehensive, and systematic methods for analyzing risk may be relatively new and rare because the processes are too novel, overly complex, and highly dynamic. As such, when data for assessing risk are insufficient, opportunitics for bias may be introduced. In these situations, Cho, Choi, and Kim (2002) suggest an inductive fuzzy approach using an event tree analysis (ETA). This model uses frequency analysis when actual data are sufficient.

When frequency data are not available, probability theories such as minimal cut sets or Bayesian approaches are used. Incomplete data can also be complemented with simulated data. In addition, if data are not available or missing, occurrence probabilities of each path can be assessed by subjective judgments based on experts' experiences and knowledge.

Even when data are sufficient, the data may not be sufficiently predictive. For example, dose-response relationships estimate the probability of an event such as cancer in an individual once exposed to a dose of some material. What happens when the data one has does not support a linear relationship between dose and response? Take the example of nitrates in drinking water. Given that dose-response relationship is nonlinear, the techniques of nonlinear fuzzy regression (Bardossy, Borgardi, & Duckstein, 1993) may be a better modeling tool.

Fuzzy models have been used for years in predicting events with uncertain, unknown, and unpredictable factors (e.g., natural disaster events such as earthquakes). Unfortunately, while there are hundreds of published papers on fuzzy systems, there is very little scholarship applied to risk communication for nanotechnology.

7.1.4. Risk Valence

The word "risk" is perceived as harmful and carries a negative valence due to notions of contamination effects. For nanotechnology, this could lead some to conclude: nanotechnology is undesirable because it involves risk. As Slovic (1986) argued, "...merely mentioning the possible adverse consequences (no matter how rare) of some product or activity could enhance their perceived likelihood and make them appear more frightening" (p. 405). Although the public seems to view contamination now as being worse than it has ever been, people today are safer than they have been at any point in history (Fitzpatrick, 2000). Yet, each generation seems to be involved in a downward spiral of risk valance in which risk is perceived to be more negative and also more plentiful. One cannot say that there is risk without negative reactions that in some cases are then exaggerated by various forms of the media (Berube, 2008c).

Policy makers have also viewed risk and cost as equivalent. For example, cost benefit analysis became prevalent with its implementation during the Kennedy Administration through the Planning, Programming, and Budgeting System (PPBS) introduced by Defense Secretary Robert McNamara and Comptroller Charles Hitch. Additionally, during the Kennedy Administration, sensitivity to technological risk rose with the adoption of nuclear energy as a peace-time power source (Short, 1984). Risk benefit analysis contributed to the creation of four major risk regulatory agencies: the **National Highway Traffic Safety Administration**[9] (1966), **the Occupational Safety and Health Administration**[10] (1970), the (1970), and the **Consumer Products Safety Commission**[12](1972). With the advent of these technological regulatory agencies, documents such as the World Medical Association's Declaration of Helsinki (1975) called for a favorable risk benefit ratio in biomedical research, and the term "cost" became increasingly interchangeable with the term "risk," lending a negative valence to its etymology.

When dealing with risk, communicators must recognize that risk has a negative valence and will not result in a positive association with whatever technology is being discussed. Consequently, it becomes important to comparatively examine risk (e.g., comparing benefits and costs) even when the benefits might be experienced by a population other than the one experiencing the costs.

7.1.5. Narratives and Anecdotes

Narratives are probably more meaningful to the public than any other strategies mentioned so far. Narratives generally take the form of anecdotes and plot-driven stories. While we recognize that the public should receive empirical data about dosage and exposure, most people experience narrative data as more accessible, approachable, and memorable. In fact, they use narrative frames when communicating what they learn to others. Additionally, people impose narratives onto data where no narrative previously existed in order to better make sense of the information (Pennington & Hastie, 1986, 1988). Narratives are also essential framing devices, which necessarily interpret information as it is being told (Faber, 2007).

Narratives have several well-researched benefits including improved memory retention (Price & Czilli, 1996), promoting knowledge integration, data organization, comprehension, addressing complex problems (Kearney, 1994), processing of information, and adding meaning to complicated and abstract information using emotion (Finucane, Alhakami, Slovic, & Johnson, 2000). In addition, Hidi & Baird (1988) showed that students were more likely to retain and recall concrete information that resonated on a personal level than they were to recall more abstract material. Satterfield, Slovic, & Gregory (2000) showed the value of a narrative-based representation of a public policy problem over a more conventional utilitarian representation.

Narratives are plot- and/or event-based (Rimmon-Kenan, 1983). These plot structures can be used in valuation contexts to assess the attributes of a problem (Satterfield et al., 2000). This in part is due to the addition of a point of view. If the audience can understand the issue through a narrator's point of view, they are more likely to feel personal agency and empowerment to take steps to solve it. Narratives also remove some of the distance between lay audiences and abstract information. If the narrative about a scientific venture and its subsequent risk profile can be created and controlled, then the public's attitude towards that venture can be informed by a more balanced counterweight as compared to the more negative and extreme perspectives stakeholders may be generating.

For many, anything at the nanoscale is abstract and for this reason, could be a good candidate for using narratives to more effectively communicate risk. For example, risk stories can become narratives in the media. This happened with the stories linking carbon nanotubes to the familiar asbestos fiber mesothelioma narrative. If narratives are only used to illustrate dangers of any new technology, such as nanotechnology, unbalanced and oftentimes negative perceptions predominate. Positive features often take the back seat to dramatic and hyperbolic tales of disaster.

7.2. New Theories and Research Findings

The following are examples of active areas of risk communication research that offer some new insights into the science of risk communication.

7.2.1. Dosage/Exposure Discounting Effect

The public does not generally know how to make sense of dosage and exposure information when understanding hazards. This is especially true for EHS issues. While experts focus on mathematical probabilities in terms of dosage and exposure, the public seems far more concerned about what the hazards are and whether they will be exposed. These differences, also known as probability neglect, result in unintended miscommunication between experts and inexperts (see section 4).

Research findings have shown that experts rationalize hazards against dosage and exposure while the public does not. Kraus et al. (1992) noted, "[t]he public would have more of an all or none view of toxicity… [T]hey appear to equate even small exposures to toxic or carcinogenic chemicals with almost certain harm" (pp. 217 & 228). This all-or-none approach seems to be taken despite well-documented hormesis effects of some chemicals when a small exposure actually triggers a beneficial effect. For example, MacGregor, Slovic and Malmfors reported "…people reserve the term 'exposure' for substantial contact or contact sufficient to cause cancer" (1999, p. 653). Consequently a fundamental variable in traditional risk assessment does not seem to find footing in public risk perception algorithms.

Weinstein (1988) when studying public sensitivity to chemicals in food, found an early example of this principle. When people were asked to respond to, "When some chemical is discovered in food, I don't want to hear statistics; I just want to know if it's dangerous or not," Weinstein elicited strong agreement from 62 percent of respondents and moderate agreement from 21.6 percent of the respondents. Weinstein's findings were supported by Kraus et al. (1992) who argued, "even a minute amount of a toxic substance in one's food will be seen as imparting toxicity to the food; any amount of carcinogenic substance will impart carcinogenicity, etc." (Kraus et al., 1992, p. 229). Evidence like this lead MacGregor, Slovic and Malmfors (1999) to conclude "...somewhat subtle changes in how the concept of exposure is conceptualized and communicated evokes very different inferences about its meaning" (p. 652). These examples demonstrate the sensitivity and need for a careful strategy when communicating to the public about these concepts.

7.2.2. Mini-Max or Maxi-Min Effects: Low probability-high Consequence Bias

The mini-max and maxi-min theorems refer to a phenomenon whereby low- probability, high-consequences events are given more attention than the scientific data supports. As an example, "To employ a mathematical analogy, we can say that although the risk of extinction may be fractional, the stake is, humanly speaking, infinite, and a fraction of infinity is still infinity" (Schell, 1982). Schell, using the example of nuclear war, illustrates how people ignore probability assessment and focus exclusively on the potential impacts.

The public seems particularly vulnerable to the maxi-min or mini-max bias (Berube, 2000). In these cases, low-probability, high-consequence events are exaggerated where public perception is heavily weighted in favor of catastrophic accidents, such as the search for the Higgs boson (dubbed the God Particle) that could produce an earth-destroying black hole (Overbye, 2008). (This event did not occur when the Cern Hadron Collider was first engaged in September 2008). This phenomenon, among others, such as the Drexler-driven nanotechnology "gray goo" scenario, has been associated with probability neglect whereby the public focuses on the worst-case scenario.

"This is largely due to news media coverage, which gives infinitely more attention to low-probability and high-consequence events than to frequently occurring, unspectacular or even undetectable events which accumulatively do much more damage to human health" (Cohen, 1985, p. 2). The consequences lead "citizens...to support expensive preventative measures, however remote the risk and however cost- ineffective the abatement procedures" (Kahan, Slovic, Braman, & Gastil, 2006, p. 1077).

> Policies driven by the consideration of low risk probabilities will, on the whole, lead to low investment strategies to prevent a hazard from being realized or to mitigate the hazard's consequences. By comparison, policies driven by the consideration of high consequences, despite low probabilities, will lead to high levels of public investment (Nehnevajsa, 1984, p. 521).

Ethical considerations that accompany exploiting low-probability, high- consequence situations for private gain or public manipulation should be carefully deliberated and weighed.

7.2.3. Rumor and Fraud

One of the most challenging research questions involving risk communication involves the intrinsic veracity of the risk information itself. This goes beyond whether the hazard or probabilities were correctly calculated. In fact, as shown below, studies have shown that rumors about risk affect behavior similarly to objectively validated risk information.

Food, cancer, and poison are considered some of the most easily spread topics for rumors—especially if they involve children. For example, a controversial study, because of the ethical issues it raised, is Kapferer's. His Villejuif food poisoning experiment involved leaving leaflets in public venues attesting to food poisoning (Kapferer, 1989). Smith (1990) claims that a researcher's failure to inform research participants about their participation in a study and exposure to false information, as in the Kapferer study, violates the Code of Professional Ethics and Standards of the American Association for Public Opinion Research.

According to Kapferer (1989), rumor research is difficult to conduct because of the boomerang effect. The boomerang effect, according to Kapferer (1990), occurs because "not only do they [warnings that the statement is not true] diffuse the rumor but they even fail to persuade those who learned it through denial" (p. 437). Nonetheless, the likelihood of the impotency of denials challenges how to ethically conduct rumor research.

Though often described as mysterious, Kapferer (1990) has claimed that rumors follow logical mechanisms. Counter-intuitive research has found that opinion leaders have a strong influence on a group and that they do not verify rumored information any more than non-opinion leaders. Although Shibutani (1966) has argued that education can moderate the effectiveness of rumors, Kapferer (1989) has disagreed, claiming an indifference to the influence of education on the dissemination of rumor.

Addressing the power of rumor in risk communication remains important since risk communicators will be faced with deciding whether rumors should be denied or not. Within the nanotechnology industry, assumptions about a contagion scenario (whereby a negative experience with a product, amplified by the media, adversely affects the product and/or industry) have yet to be proven.

7.2.4. Risk Profile Shifts

When communicating risk information, individuals and groups can and will use different rhetorical devices. One such device is a "risk profile shift" (Berube, 2008a).

In risk précis, each event or phenomenon is attributed a certain risk profile. This profile represents a composite of estimates of the degree of hazard. The degree of hazard is modified by dosage variables based on research data and by exposure variables often expressed as probabilities. Each type of nanoparticle has a specific risk profile. However, so does a chair, an automobile, or an oven.

Usually an event or phenomenon belongs to a specific class or set of phenomena. Risk profile shifting occurs when the risk profile of one particular phenomenon in a class of phenomena is transferred to a different phenomenon within the same class. The acceptability of such a transfer is a function of the perceived homogeneity of the class from which the phenomenon was drawn. The more heterogeneous the class the higher the likelihood that the transference is invalid. An example of a risk profile shift occurred in a recent report by Friends of the Earth on sunscreens, and is described in detail by Berube (2008a).

Risk profile shifting becomes a rhetorical device when people over-generalize and misattribute the risk profile of a more benign phenomenon to one that is more malignant or vice versa. This misattribution may be due to personal or organizational intentions, ignorance, or interpretation differences. In general, this practice should be considered unethical because it spreads misinformation, usually to inexperts and publics with little to no expertise or background in science. Risk profile shifts also illustrate the need to vet information in science and technology more broadly and from all sources as it enters the public sphere.

7.2.5. Power Language

Rhetorical flourishes as power language raise issues in risk communication. Richard Weaver (1953) described the powerful influence of words he called "God" and "devil" terms. In America, for example, "liberty" is a God term while "terrorism" is a devil term. When a risk communicator remarks: "We don't need another 'Frankenfood' disaster!" he or she is attempting to transfer the negative valence of "Frankenfood" (a euphemism for genetically modified foods) to the subject at hand. In this illustration, "Frankenfood" functions as both a devil and charismatic term as it carries considerable imaginative power. Weaver defined charismatic terms as "terms of considerable potency whose referents it is virtually impossible to discover or to construct through imagination." (227). Charismatic terms are assigned a socio-cultural and oftentimes political valence as well:

> As in the case with charismatic authority, where the populace gives the leader a power which can by no means be explained through his personal attributes, and permits him to use it effectively and even arrogantly, the charismatic term is given its load of impulsion without reference, and it functions by convention. The number of such terms is small in any one period, but they are perhaps the most efficacious terms of all (Weaver, 1953, pp. 227-228).

For example, if a communicator links research on carbon nanotubes and mesothelioma with repeated references to asbestos, power language using a charismatic term is attempting to transfer and associate the negative valence of asbestos to carbon nanotubes. Charismatic terms are highly rhetorical and powerfully persuasive. They function heuristically, providing the most naïve audiences with a mental short cut to assigning meaning to a subject field about which they know very little and may not want to learn more. This phenomenon bespeaks the need for experts and critics to debunk overclaims and rhetorical flourishes that muddy public perception of risk.

7.2.6. Risk Hormesis

Hormesis is the tendency of biological organisms to react favorably to lower exposures of materials that would prove harmful at higher dosages (e.g., botulinum toxin is commonly used in cosmetic operations). Hypothetically, some toxins may have a hormetic effect by inducing positive reactions in the body due to the stimulation of biological repair functions that respond to a higher level of threat than is actually present.

The "contamination effect" is the belief that once an individual is exposed to a substance he or she is "contaminated" for life: once in contact, always in contact. For example, Slovic (1998) cites surveys that show very large percentages of the public believe that if one is even exposed to a carcinogenic chemical once, he or she is likely to get cancer some day. Three quarters of the population surveyed indicated that if even a tiny amount of a carcinogenic chemical is found in tap water, they would not drink it.

Renn (2003) argues that the hormetic effect should be communicated to the public in risk messages thereby enhancing competence and trustworthiness of the source. Positive effects of some chemicals can be informative and offer guidance to designing risk communication messages. For example, exposure to low doses of certain carcinogenic chemicals may be beneficial to a large majority of people, though a small portion of the population might be highly sensitive and experience negative effects. Maybe this is why naturally occurring nanoparticles to which we have been exposed for centuries do not seem to carry a significant EHS footprint. He argues that

> This calls for a continuous evaluation program. In particular, one needs to monitor public perception of risk, make a strong effort to collect and process public concerns and to make sure that all communication material is designed to meet public demands. This is particularly necessary in the case of hormesis, since all perceptive mechanisms imply a skeptical to negative predisposition toward making use of this theory in practical management decisions (Renn, 2003, p. 23).

Toxicological hormesis, despite its controversial nature, may be useful in risk communication as a metaphorical concept. Risk hormesis as a metaphor is similar to the anchor heuristic (see Table 5); if presented with a little risk, the public will be more likely to accurately evaluate the potential for larger risks and less likely to blindly adopt more extreme information. Prompting a moderate amount of critical thought about minor risks could lead to the same critical thought being applied to greater possibilities for risk. This strategy, called the "inoculation effect," has been validated in political campaign studies for decades (Pfau et al., 1990). This would, in effect, place the public in a similar position as the scientists, for scientists are required to evaluate all degrees of risk from minor to severe. Such a risk-hormetic effect may also explain why scientists tend to view risk differently than the public does.

7.2.7. Risk Homeostasis or Risk Compensation

The theory of risk homeostasis hypothesizes that an individual has a built-in target level of acceptable risk that fluctuates around a set point (much like how temperature is regulated by a thermostat in a home) (Wilde, 1982; 2001). This set point varies between individuals. The theory of risk homeostasis provides a relative range of behaviors and is not predictive of specific human behavior in all situations. "It would be foolish to think people never adjust their behavior in response to perceived risks, but a more productive approach is to try to determine the conditions under which this occurs" (O'Neill & Williams, 1998, p. 93).

Peltzman (1975) reported evidence of people driving faster and less carefully when safety improvements are added to cars and trucks. Peltzman compared pre- regulation and regulated periods and concluded that newly legislated vehicle manufacturing standards had not led to a reduction in highway death rates. Wilde (1984, 2001) has suggested that when a level of acceptable risks in one part of an individual's life changes, there is a counter-corresponding rise/drop in acceptable risk elsewhere for individuals and larger human systems (e.g., a population of drivers).

Peltzman and Wilde's theories have attracted criticism, especially conclusions regarding driving behavior (see O'Neill & Williams, 1998). While Evans and Graham (1991) state that Wilde's theory commands as much credence as the flat earth hypothesis, this hyperbole conflates individual and population levels of risk homeostasis and offers little insight into why the data sets allow competing conclusions.

Risk homeostasis has many implications. It may help to explain the popularity of such practices as gambling, frightening amusement park rides, high-speed driving, sky- diving, etc., and may also explain that as society becomes safer there is a corresponding increase in risky behavior. Simply put, people are cognitively designed to deal with and respond to risk. Once outlets for response are closed, people find other ways to behave to maintain a balance of risk in our lives.

7.2.8. Risk Fatigue

Exposure to risk messages matter. As members of the public become more overwhelmed with risk messages, they attend to fewer of them. Unfortunately, when they do pay attention, they attend to the wrong risks as well. Thus, the proliferation of risk messages has produced risk fatigue. According to Professor of Geography John Adams from University College London, "risk fatigue" is a state of cynicism engendered by the popular media's habit of sensationalizing every newly discovered virtual risk (Adams, unpublished work). Risk fatigue is "a general air of cynicism because scientists say one week that something is dangerous and a few weeks later they contradict themselves" (Blakemore, 2000). Risk fatigue is an important and potentially dangerous problem that should be considered in communication and engagement strategies. A classic illustration of this is the rhetoric associated with terrorist alerts. While we know there have been alerts and some know there is a color-coded risk ranking system, few people can explain what each color in the risk system means. Much like the donor who has received too many requests for her philanthropy, the public has been exposed to too many risk messages and is experiencing risk fatigue.

Eckersley (2001) used the term "risk fatigue" when he addressed obesity issues in Australia.

> Health promotion campaigns are underestimating the social determinants of health, and "risk fatigue" is affecting attitudes to complying with healthy lifestyle standards.... When it comes to reducing risk behaviors and promoting healthier lifestyles, there is a real danger of causing "risk fatigue" by asking too much of people too often.

According to the UK's Food Standards Agency Executive Director Patrick Wall, some food-related incidents were left out of the European Union's White Paper on Food Safety (Commission 2000) because, "Consumers were showing signs of 'risk fatigue' in the face of messages that understandably could be interpreted as taking the joy out of eating" (See O'Sullivan, 2000). Jones also showed that meat consumption was related to publicized scares (consumption of meat products dropped in the mid-nineties but eventually recovered) and argued that over-reporting might have produced a fatigue that helped normalize consumption levels and that "[t]o some extent the public seems to be suffering from 'health risk fatigue' and consumption has recovered about half of the initial loss with low prices helping consumption to recover."

While individuals in industrial countries are enjoying greater health and longevity, the amount of analysis of risk is growing at an increasing rate. Critics are expressing the fear that people are developing risk fatigue and blaming scientists for this over- production of risk.

7.3. Summary and Conclusion

This section served to briefly address major research frameworks, communication design considerations, and future research directions on risk assessment and risk communication. This information should inform risk communication strategies. For example, it will be important to consider the role of fatigue when generating risk-related messages; one good message will probably trump a dozen weaker messages. Finally, issues of how to deal with counter messages that involve risk profile shifts or that are heavily laden with power language will also need to be addressed.

SECTION 8. RECOMMENDATIONS

The following is a risk communication algorithm that serves as an organizing structure for understanding factors and theories about risk perception and communication theory. To illustrate this algorithm, we have used the following risk communication model as a way of illustrating how to improve communication between expert and inexpert audiences.

8.1. Prevailing Methodologies

There are three ways to deal with the current shortcomings of risk communication about emerging technologies, especially nanotechnology.

- We can leave risk communication to public relations and marketing experts. As they generally come to the subject after the fact to deal with a crisis or catastrophe, this approach does little to support goals of proactive participation and engagement.
- We can leave science and engineering to scientists and engineers and risk communication to risk communicators. While this approach may initially be appealing, there are not enough qualified professional risk communicators to cover the subject field and some communication contexts—like science to media—may benefit from technical input.
- Experts, including scientists, engineers, and policymakers, need a basic understanding and skill set to increase their effectiveness when communicating risk to the media and the public.

Using the third approach, developing a basic skill set for scientists, engineers, and policymakers begins with re-examining an algorithm for communicating risk: Risk ← Hazard (Exposure + Dose) that, as mentioned earlier, may not be adequate when addressing dosage, hazard, and exposure variables, especially given the paucity of quantitative values for hazards as loss of privacy and technological prejudice.

A competing algorithm was introduced by Sandman: Risk ← Hazard + Outrage. While Sandman's algorithm makes the case that for members of the public, their determinations of risk ignore probability for other variables. Shortcomings to Sandman's algorithm include not distinguishing between technical and public audiences and not covering the breadth of variables about which the public is preoccupied.

8.2. A New Risk Communication Algorithm

As an exercise, we are offering the following algorithm as an alternative to other risk algorithms in use today:

$$\text{Risk} \leftarrow \frac{\text{med}(Ev_1 + Ev_2 + \dots Ev_n)\ \text{med}(P_1 + P_2 + \dots P_n)}{T} \quad I$$

Ev = event, **P** = probability, **I** = information, and **T** = trust.

Ev: An event can have either a positive or negative valence. Estimates for event instant valences range from 0 →1, with 0.5 being a hazard-neutral event and we use the median value rather than the mean.

P: is a function of exposure and dosage. Its estimated values range from 0 →1, where 0.5 describes a random occurrence and we use the median value as well.

We chose **med**ian value because it would exclude outlier scenarios and estimates since oftentimes they are maxi-min phenomena. For example, CERN's large Hadron collider scenario is a classic illustration of why it is important to use median rather than mean estimates.

Lederman and other physicists refer to the theoretical Higgs boson particle as the God particle because its discovery would lead to the unification of many disparate particle physics theories and a plethora of discoveries that could redefine and change life as we know it. Others, as mentioned earlier, warn that as the collider searches for the God particle it may at the same time generate micro black holes and strangelets that could doom the planet, if not the universe.

Both the positive and negative claims carry such a high valence with infinity or near infinity as a value, that any probability estimate is impossible.

I: Information includes all data introduced by individual(s) and groups whether mediated or not. An assumption is made that the media will moderate data with media- generated source credibility values (a trust-like variable that can be different from public source trust). Fear is a public information variable that is brought to the event by the public. Fear is affected by outrage to the event and stigma, an event category media- amplified emotive response, most often associated with affected populations. Both of these tend to be highly specific to the event. Dread is viewed as an emotive response to the interplay between outrage and stigma. A value of 0 for **I** would describe an unreported risk event and an event instant without any risk apprehension, hence no risk event.

T: Trust is used as source trust and is a public variable that moderates the fear the public brings to the event. A value of 0 would describe total distrust in the source of information. A value of 1 would describe total trust and truth.

Codicils:

1. For purposes of simplicity, positive values ranging from 0 to 1 are used.
2. Event rather than hazard as a variable was used to compensate for the negative valence associated with hazard and to return benefits to the risk algorithm to support comparative risk communication.

8.3. A New Risk Communication Metaphor

Education research has shown that one of the more effective ways of introducing complex concepts to novice audiences is through metaphors. When dealing with inexpert audiences, it is fortuitous to design links between known events and unknown ones and that is accomplished with metaphors and other imagery. The following metaphor was designed to help experts and decision makers communicate complicated risk communication concepts to inexpert audiences. It is only a working metaphor and a tool.

The following example of auto racing and the speedway will illustrate the utility of this approach.

AGENTS: Who and what they are. Many people and objects are involved in a race and they have certain roles much like the many roles played by those who generate and encode messages as well as the channels involved.

Announcers: Traditional media.

The announcers call the events of the race for the benefit of the spectators (public). The announcers frame the race using a traditional broadcast format. Another type of framing involves the announcers' choice of commentary.

Crew: Risk communicators.

Individuals involved in communicating risk, who are represented by the crew, help to assess and determine whether a car (message) can withstand the turmoil of a race. A good crew should be able to spot weaknesses in a car and be able to make the necessary adjustments to ensure that a particular car is the focus of the spectators (public). The crew often assists in maintaining the car during the race, thus helping to keep it safe on the racetrack. In addition, the crew is often guided by the Crew Chief.

Crew Chief: Professional risk communicators.

The Crew Chiefs are the foremen of the risk communication staff and in charge of the car (message), mechanical crew (risk communicators), and their choice of operations. The Crew Chief is an important leader because he communicates directly with the drivers (sources of information) and other crew members (risk communicators). The Crew Chief takes control of the situation and should be alert to all aspects of the race. The Crew Chief makes strategic decisions about the course a racecar should take and it is also the Crew Chief who articulates to the crew any desired changes in the car.

Drivers: Sources of Information.

Risk communication begins with the encoding of the message and the ongoing refinement of it as well. In this metaphor, racecar drivers are equivalent to the entities that investigate risk and create messages to be disseminated to the public. These sources of information can be viewed as four "racing teams" (represented by their cars' team colors). They are:

- Academic scientists,
- Government reports,
- Industry/Marketing/Insurance, and
- Non-governmental organizations.

In the racing world, drivers are usually supported by particular products/companies typified by graphic advertisements. These companies would like to see their car and driver win the race because that car and driver will be the focus of the media and most salient to the audience (public). Similarly, in the world of risk communication, there is competition between groups to produce the most powerful and salient messages to be disseminated to the public. In addition, drivers fine tune the message when they make decisions behind the wheel.

Racecars: Messages driven by sources of information.

This model represents racecars as the combination of data and information provided by the source. This includes the way information is framed and disseminated by risk communicators and the media. In short, the racecars represent risk messages presented to the public. All race cars are not alike as all messages are not alike. Some are designed and built better than others, as are risk messages.

Emergency Medical Technicians: Crisis communicators.

Once racecars have crashed (messages have spun out of control) the ambulance (crisis communicators) jumps into action to help get the situation back on track. The EMTs have to work to repair driver injuries (source) and provide the spectators (the public) with a sense that the situation is being handled expertly and efficiently.

Grandstand spectators: The public.

The spectators in the grandstands represent members of the general public who consume the information they are provided. They do not generate information per se. They pay attention to the racecars on the track but are not often as concerned about the inner workings of the track or the sources of information (drivers). Their main concern is the interplay of cars (messages) and which car wins the race (the message the public will accept).

Infield spectators: Digital media.

The infield spectators are very enthusiastic characters in the race. Like the grandstand spectators, the infield spectators represent the public. However, because of their enthusiasm, these spectators are more involved in viewing the outcome of the race. They may have their own opinions of the events and the final turnout. They may voice these opinions (new media production and sharing) before, during, and after to fellow fans of the race. There are over

38,000 hits in Technorati of blogs related to NASCAR and some infield spectators web-broadcast during the races.

Race officials: Policy makers.

The race officials declare the winner of the race (particular message focused on by the public) but are not involved during the race or determine the path of the cars (messages). They decide what happens to the cars during and at the end of the race with rules and flags. If the race officials declare a car the winner, this is similar to enacting public policy according to a particular winning risk message. Winning is the first finisher of the race, the car and driver that the spectators (public) sees as the best.

Speedway: Paths or channels that messages take.

Just as the racecar track changes, so do situations in risk communication. The track is neither linear nor a perfect circle and what happens to the course of the car (message) is never wholly predictable. In addition, part of the track is preferred. It is called the groove and is the best route around a race track (see avoiding over- and under-steering). Collisions with other vehicles, spin-outs, penalties, and race officials (see below for comparisons) all influence the final outcome of the race or what happens to the risk message.

ACTS: What they do. Here are some activities that can be found during a race and descriptions of what the various agents do under these circumstances.

Advertisements on cars: Static.

The advertisements are important to various groups (sources) supporting the race but they contribute little to the cars (messages) and the paths taken in the race. However, they are effective in team building since cars with the same advertisements often are on the same team. Some advertisements are larger and glitzier than others. Messages can be festooned with all the bells and whistles associated with public relations, but risk communication has a deeper character than simply marketing messages and a poorly designed message dressed up will have little long-term effect.

Pace Car lap: Attenuation.

Sometimes, prior to a crash or following a crash, a yellow flag is waved by the race officials (policy makers) signifying the arrival of the Pace Car. The Pace Car drives around the track so that the cars slow down. This is similar to attenuators slowing down a message. This also makes the race less exciting and the spectators (public) are not as concerned with the activities of the cars (risk message).

Pile-up: Amplification scenario.

In a pile-up one car causes crashes involving cars immediately following it. Sometimes a poorly designed risk communication message can cause more problems. Racecars crashing (messages collapsing or getting out of hand) can produce an emergency situation.

Staggering the start/pole positioning: Presumptive credibility.

When the race starts, all of the cars simply cannot fit in the first row. As such, they are staggered. This is not a random exercise but the product of qualifying procedures. The best performing drivers in the qualifying laps are like messages involving higher degrees of trust and are consistent with values the public already hold true.

AGENCY: How they do it. These are choices drivers make to improve their overall performance.

Avoiding over- and under-steering: Risk communication training.

The driver would like to stay in the groove to avoid over- and under-steering. There are many principles associated with risk communication that have been gleaned from previous experience. These errors need to be avoided as do over- and under- steering in a race.

Backing out/dropping the hammer: Responding and rebuttal.

Backing out refers to slowing down while dropping the hammer involves full acceleration sometimes referred to as "on the throttle." Some messages involve denial and rejection while others are much more aggressive and part of a vibrant campaign. Risk communication messages have a velocity and intensity of their own and risk communication can adjust the velocity and intensity of the messages to produce the most effective results.

Blocking: Inoculation.

If there are multiple messages competing for public attention, this is the practice of one risk communicator preventing others from getting as much attention, if any attention at all, by anchoring public perception more positively.

Bump-draft: Priming.

A bump draft places cars really close to one another and actually pushes the front car. This makes both or several cars move faster, but can be very dangerous when cars spin out of control. This concept is similar to framing or agenda setting because sources can change a message so that it is picked up by the media more readily and moves more quickly to the public. In some instance, poorly designed priming can be counterproductive.

Drafting: Transference.

When the cars are jockeying for position, cars will "draft" behind another. A "draft" refers to a gap in air currents left in the wake of a vehicle traveling at high speeds. When another racecar gets closer to the racecar in front of it, it faces much less wind resistance while in the "draft" of the car ahead of it. When one message is relying on the success of the message preceding it (anchoring), it transfers credibility and other features (trust) to the car (message) that is drafting.

Holding up traffic: Social cascade.

This term refers to a driver who purposefully causes cars running faster on the track to slow. In this instance, the driver manipulates the pace for strategic reasons much like a risk communicator. Oftentimes the amplification surrounding a hazard event needs to be slowed down sufficiently to reduce the likelihood of a social cascade or moral panic. Attenuation is most effective if the pace of some of the messages can be slowed.

Hole shot: Tactical response.

This term in speedway racing refers to finding an opening and shooting through the opening before opposing drivers can react. In risk communication, it can refer to taking full advantage of a break in the battery of the messages.

Inside and outside grooves: Case/client-specific campaign.

While the part of the track that is preferred is called the groove and for most everyone is the best route around a racetrack, there are some exceptions. The "high groove" takes a car closer to the outside wall for most of a lap. The "low groove" takes a car closer to the apron than the outside wall. Sometimes a car will handle and perform better on the outside or inside line and a driver opts not to use the optimum groove. Different drivers and different racecars perform differently on different grooves and during different parts of the race drivers search different grooves for different purposes. This can refer to the tuning of a risk communication campaign to fit the specific needs of the moment and the client.

Motoring: Trust building.

When a driver is using the racecar in a prudent and wise fashion and not demanding more of the car than it can perform, this is called motoring. This can refer to the efforts to maintain ongoing high levels of trust through transparency and public engagement.

Sandbagging versus going “flat out”: Strategic argumentation.

In the first situation, the driver allegedly fails to drive a car to its full potential in practice or qualifying, thus being able to provide a "surprise" for competitors during a race. In going “flat out” the driver attempts to get his racecar to perform at its highest efficiency. The velocity of the risk communication campaign is determined tactically. Sometimes you need to hold you best argument for the end of the campaign while at other times you put the “pedal to the metal.”

Wedge and weight jacking: Tactical argumentation.

The crew gets to fine tune the racecars. These two techniques of many involve improving performance by modification. Wedge or bite involves the adjusting of car-jacking screws found on each wheel. This weight jacking redistributes the car’s weight at each wheel and affects performance positively.

OTHER: Support systems.

Parade lap: Pre-testing methodologies.

The warm-up lap before a race is the parade lap. Drivers use this lap to warm up their engines and often zigzag to warm up tires. In risk communication this is akin to the use of focus groups to test messages before they are released.

Pit board: Continuous input/output.

A board used by crews to inform drivers of lap times, lap until pit, and other various bits of information. The board is used along with team radios to keep in constant communication. This racing term emphasizes the need for continuous communication between those soliciting risk communication and those who craft it. The alternative has generally been to wait until a crisis and then call in a crisis communicator. This practice is hardly painless for all involved. While you might leave risk communication to professional risk communicators, you should expect them to maintain communication with you.

Racer's tape: Quick fix.

Heavy duty duct tape (usually gray) used to temporarily repair hanging body parts, which might hinder aerodynamic features and decrease performance. This resembles the quick fixes often used by crisis communicators to correct for particularly egregious problems that result from a failed attempt by an inexperienced communicator, which only worsened the risk communication situation.

Programs, fan events, etc.: Ongoing public engagement.

There are programs and brochures. There are pre-race exhibits, lectures, and classes to better enjoy the racing experience. There are fan clubs, auto shows, and special events, like the Stock Car Racing Fan Fests, where the public meets the drivers and sees the racecars up close. All of these events are similar to exercises in public engagement. By providing some information and access to the drivers and others associated with the racing event, the public better understands what is happening and enjoys the event.

Codicils:

1. There are many varieties of auto racing and this metaphor does not refer to any specific league of racing.
2. Though critics may think an auto racing metaphor is too male oriented, this sentiment is inaccurate. In NASCAR for example, female fans now comprise 40 percent of the fan base and half of its new fans according to 2006 Nielson survey. Furthermore, the number of female drivers is growing in all leagues of auto racing especially after the success of the Indy Racing League's Danica Patrick and Milka Duno as well as many others.

Discussion: No metaphor is perfect but this metaphor is useful because of the strong connection between drivers (sources of information) the crew (risk communicators), and spectators (the public). If people can understand how important the crew is to the success of the race, they can also begin to understand the importance of risk communicators in determining the path and success of risk messages.

8.4. Some Directions for Future Research

After an extensive review of the literature, we note the following areas as research-appropriate. While we will consider work in these areas, we hope others will as well.

8.4.1. Risk and Morbidity

We noted that too much of the literature has fixated on mortality data in establishing hazard and risk variables. While some have studied impacts from acute versus chronic exposures to chemicals, few have considered how the public rationalizes threats leading to long-term morbidity. Injury- and disability-related safety and health concerns can outweigh end-of-life considerations under some circumstances. We sense that the public suffers from morbidity neglect and may fail to consider the true value of hazards that do not impact mortality directly.

8.4.2. Axiologies and Belief Structures

We have barely skimmed the surface of the role belief structures may play in forthcoming debates over applied nanoscience. Issues associated with fertility, contraception, stem cells, and a range of nanobiotechnology and synthetic biology developments may tap religious sensibilities and belief structures (Scheufele & Brossard, 2008). In persuasion, debunking or rebutting a belief-based argument is challenging. We sense that as the nano-bio interface becomes more prominent we can expect more faith-based challenges to applied nanoscience as more belief structures concerns enter the arena of societal issues. Axiologies are complex web-like structures that affect how values are built and sustained. Just as it is extremely difficult to convert a devotee from one religion to another, it may be equally difficult to affect perceptions drawn from warrants based on belief.

8.4.3. The "Big Sort" and Risk Maps

In two recent books, Richard Florida (2008) and Bill Bishop (2008) argue that the American public is increasingly forming concentrated communities with similar opinions, interests, and perceptions. While this theory may need to be re-examined following the national election in 2008, there is much to this concept worth exploration. The "Big Sort" theory contrasts selected likeness to a vision of a more generally heterogeneous landscape of American opinion evenly distributed across the country. The attraction of like-minded individuals to communities creates a positive feedback loop in which scaffolding of large bases of support lead to amplifications of opinions. Thus, as time proceeds these communities not only could become denser with like-minded individuals but more extreme in the directionality of their views. This understanding might assist us in designing risk maps (Stone, 2001; Stoffle & Stone, 1991) appropriate for broader concepts such as emerging technologies.

8.4.5. Digital media and crises

Do people turn to digital media in crisis situations? There is strong evidence that digital media was especially important during three recent disaster events: the terrorist attacks during September 2001, the 2004 Southeast Asian tsunami, and the Gulf Coast floods during Hurricane Katrina in 2005, when traditional media fell far short of needs and expectations. We have learned that some web portals experienced much more traffic than others. Chat rooms were particularly powerful communication channels as well. During the May 2008 devastation of Myanmar by Cyclone Nargis and the August 2008 crackdown by China on Tibetan monks, data on these events were restricted by officials from Myanmar and China who closed down digital media sites as well as restricted physical access to these events. As such digital media have become a powerful tool in the management of public opinion.

Public engagement must serve a function beyond checking off a box on a grant application. If the goal of public engagement is democratic participation, we may need new models. We may also need new models if the goal of public engagement is to affect opinions by group opinion leaders. If the goals are to test whether discussion and debate in a controlled setting can affect opinion or to build trust, we can remain with the public engagement models we have today.

The public sphere has flattened with the onset of digital media. Opinion leaders are neither solely white landowners nor elected or appointed officials. Sometimes they are media celebrities, or at other times complete unknowns who quickly rise to prominence and then disappear from the scene entirely.

Digital media has changed how the public experiences events in science and technology as well as other subject fields. As Generations X and Y become policy makers, the sources of their information and many other salient and exigent variables about the digital media they favor will become more important.

Digital media is an important variable in all communication, especially disaster event driven communication, and deserves serious attention in risk communication research in the current decade.

Given the trends toward Internet and social media as primary sources for information of all sorts, including information on health and safety, debates over nanotechnology will draw from these sources for backing and claims. Indeed, as nanotechnology makes more inroads onto the market in the fields of medicine and health, energy production and efficiency, and food production and safety, the inexpert audiences, the general public, will find arguments in the world of digital media rather than on more traditional media, such as newspapers and television.

Governments and industries are making their presence known on web sites. Some have had the foresight to utilize the resources of social media to market their ideas and products and as a platform for public engagement. Risk communication will demand a strong and vibrant relationship with digital media to be effective in the 21st century.

APPENDIX A

Reading List on Risk Communication

The following three books are artifacts in risk communication. They can be read to understand the developments in the field. The second set of readings below is recommended.

National Research Council. (1989). *Improving Risk Communication.* Committee on Risk Perception and Communication, Comission on Physical Sciences, Mathematics, and Resources. Washington D.C.: National Academy Press.

Committee on the Institutional Means for Assessment of Risks to Public Health, Commission on Life Sciences National Research Council. (1983). *Risk Assessment in the Federal Government: Managing the Process.* Washington D.C.: National Academy Press.

Sandman, P. M. (1993). *Responding to Community Outrage.* Fairfax, VA: American Industrial Hygiene Association.

Recommended Readings

Adam, B, Beck, U. and Van Loon, J. (Eds). (2000). *The Risk Society and Beyond: Critical Issues for Social Theory*. London, UK: Sage Publications.
This edited volume includes twelve essays on a wide variety of topics including biotechnology and human enhancement. The conceptual articles in parts one and four, especially Scott, Lash, Rose, and Beck, are especially worthwhile. Berube has used this book in his graduate seminar in risk studies as a discussion vehicle.

Breyer, S. B. (1993). *Breaking the Vicious Circle: Toward Effective Risk Regulation.* Cambridge, MA: Harvard University Press.
This short book discusses opportunities to regulate risk through conventional risk governance models. Berube recommends this book to students in public administration and policy science.

Gardner, D. (2008). *Risk: The Science and Politics of Fear.* London, UK: Virgin Books. Most bookstores do not carry this book and this is mostly due to the publisher. Gardner provides an accessible examination of risk as perceived by experts and the inexpert and is a solid introduction to public heuristics and biases. Berube recommends this book to anyone with an interest in the field of risk perception and communication.

Lundgren, R. E. & McMakin, A. H. (2004). *Risk Communication: A Handbook for Communicating Environmental, Safety, and Health Risks.* Columbus, OH: Battelle Press.
This is an undergraduate textbook for a risk communication class. It examines risk communication and provides useful information and recommendations. Berube recommends this text to business communicators with an interest in broadening their knowledge base about risk communication.

National Research Council. (2008). *Public Participation in Environmental Assessment and Decision Making.* T. Dietz and P. Stern. (Eds). Committee on the Human Dimensions of Global Change, Division of Behavioral and Social Sciences and Education. Washington, DC: National Academies Press.Prepublication Copy. Uncorrected Proofs.
This report is not only very new but also read as an uncorrected proof. While this report is specific to public participation and environmental issues, The report's third conclusion is much like ours: "design it to address the challenges that arise from particular contexts.... There is no single best format or set of procedures for achiving good outcomes in all situations." Unfortunately, the report defers to the 1996 publication *Understanding Risk* by Stern and Fineberg that doesn't focus on communication and offers little guidance to the challenges of digital media.

Sunstein, C. R. (2002). *Risk and Reason: Safety, Law, and the Environment.* Cambridge, MA: Cambridge University Press.
This book does a highly credible job introducing the reader to risk and how it interplays with the precautionary principle. It is well written and strongly recommended to students in public administration and policy science who want to understand the full range and significance of the precautionary principle. Berube highly recommends almost everything Sunstein, an economist from University of Chicago, has written in the area of risk.

APPENDIX B

Bibliography of Research on Public Perceptions of Nanotechnology

Bainbridge, W. S. (2002). Public attitudes toward nanotechnology. *Journal of Nanoparticle Research*, *4*(6), 561-570. DOI: 10.1023/A:1022805516652

Besley, J. C., Kramer, V. L. & Priest, S. H. (2008). Expert opinion on nanotechnology: Risks, benefits, and regulation. *Journal of Nanoparticle Research*, *10*(4), 549- 558. DOI: 10.1007/s11051-007-9323-6

Bundesinstitut für Risikobewertung. (2006). Consumer Conference on Nanotechnology in Foods, Cosmetics and Textiles. November 20. http://www.bfr.bund.de/cm/245/bfr_consumer_conference_on_nanotechnology_i n_foods_cosmetics_and_textiles.pdf. Accessed February 28, 2009.

Bundesinstitut für Risikobewertung. (2007). Majority of Consumers View the Development of Nanotechnology Favourably. Press Release 23. December 19. http://www.bfr.bund.de/cd/10563. Accessed February 28, 2009.

BMRB Social Research. (2004). Nanotechnology: Views of the general public, quantitative and qualitative research carried out as part of the nanotechnology study. *BMRB International Report 45101666*. http://www.nanotec.org.uk/Market%20Research.pdf. Accessed June 30, 2008.

Burri, R. V. & Belluci, S. (2008). Public perception of nanotechnology. *Journal of Nanoparticle Research*. *10*(3), 387-391. DOI: 10.1007/s11051-007-9286-7

Cobb, M. D. (2005). Framing effects on public opinion about nanotechnology. *Science Communication*, *27*(2), 221-239. DOI: 10.1177/1075547005281473

Cook, A. J. & Fairweather, J. R. (2007). Intentions of New Zealanders to purchase lamb or beef made using nanotechnology, *British Food Journal*, *109*(9), 675-688. DOI: 10.1108/00070700710780670

Currall, S. C., King, E. B., Lane, N., Madera, J. & Turner, S. (2006). What drives public acceptance of nanotechnology? *Nature Nanotechnology*, 1, 153-155. DOI: 10.1038/nnano.2006.155

Danish Board of Technology. (2004). Citizens′ Attitudes towards Nanotechnology. June. http://www.tekno.dk/subpage.php3?article=1093&language=uk&category=11&toppic=kategori11. Accessed February 28, 2009.

Einsiedel, E. (2005). In the Public Eye: The Early Landscape of Nanotechnologies among Canadian and US Publics. *Journal of Nanotechnology Online*, 1, December. http://www.azonano.com/Details.asp?ArticleID=1468. Accessed February 28, 2009.

European Commission. (2001). *Europeans, Science and Technology*. Eurobarometer Special Survey 154, organized and supervised by DG Press and Communication, Brussels.

European Commission. (2005). *Europeans, Science and Technology*. Eurobarometer Special Survey 224, organized and supervised by DG Press and Communication, Brussels.

Gaskell, G., Eyck, T. A., Jackson, J. & Veltri, G. (2004). Public attitudes to nanotechnology in Europe and the United States. *Nature Materials*, 3, 496. DOI: 10.1038/nmat1181

Gaskell, G., Eyck, T. T., Jackson, J. & Veltri, G. (2005). Imagining nanotechnology: Cultural support for technological innovation in Europe and the United States, *Public Understanding of Science*, *14*(1), 81-90. DOI: 10.1177/0963662505048949

Gavelin, K., Wilson, R. and Doubleday, R. (2007). Democratic Technologies? The Final Report of the Nanotechnology Engagement Group (NEG). Involve 2007, http://www.nanowerk.com/nanotechnology/reports/reportpdf/report105.pdf. Accessed February 28, 2009.

Hanssen, L. & van Est, R. (2004). The Double Message of Nanotechnology: Research into Rising Public Perceptions. Den Haag, Rathenau Institute, October. http://www.rathenau.nl/showpage.asp?item=698. Accessed February 28, 2009.

ISTPP 2003 National Nanotechnology Survey. College Station, TX: Institute for Science, Technology and Public Policy, George Bush School of Government and Public Service, Texas A&M University, (June 16, 2004), unpublished.

Kleinman, D. & Powell, M. (2005). Madison Area Citizen Consensus Conference on Nanotechnology, Madison, WI. April 24. University of Wisconsin. http://www.nanocafes.org/files/consensus_conference_report.pdf. Accessed February 28, 2009.

Komm-passion Group. (2005). Wissen und Einstellungen zur Nanotechnologie, Frankfurt. See Schutz, H. & Wiedeman, P. M. (2008).

Lee, C. J. & Scheufele, D. A. (2006). The influence of knowledge and deference toward scientific authority: A media effects model for public attitudes toward nanotechnology. *Journalism & Mass Communication Quarterly*, 83(4), 819-834.

Macoubrie, J. (2006). Nanotechnology: Public concerns, reasoning, and trust in government. *Public Understanding of Science*, *15*(2), 221-241. DOI: 10.1177/0963662506056993

MAST. (1989). Policy Research Project on Anticipating the Effects of New Technologies, *Assessing Molecular and Atomic Scale Technologies (MAST)*. Austin, TX: Lyndon B. Johnson School of Public Affairs, The University of Texas at Austin.

Nanosciences. (2007). Dialogues avec le CNRS, April-June, http://www.insp.jussieu.fr/grand_public/debats_nano.htm. Accessed February 28, 2009.

Nerlich, B., Clarke, D. D. & Ulph, F. (2007). Risks and benefits of nanotechnology: How young adults perceive possible advances in nanomedicine compared with conventional treatments. *Health Risk & Society*, *9*(2), 159-171. DOI: 10.1080/13698570701306856

Peter, D. Hart Research Associates. (2006). Public awareness of nano grows – majority remain unaware. *The Woodrow Wilson International Center for Scholars Project on Emerging Technologies*. http://nanotechproject.org/78/public-awareness-of-nano-grows-but-majority -unaware. Accessed October 3, 2006.

Peter, D. Hart Research Associates. (2007). Awareness of and attitude toward nanotechnology and federal regulatory agencies. http://www.nanotechproject.org/138/9252007-poll-reveals-public-awareness-of-nanotech-stuck-at-low-level. Accessed October 10, 2007.

Scheufele, D. A., Corley, E. A., Dunwoody, S., Shih, T.-J., Hillback, E. & Guston, D. H. (2007). Scientists worry about some risks more than the public. *Nature Nanotechnology*, *2*, 732-734. DOI: 10.1038/nnano.2007.392

Scheufele, D. A & Lewenstein, B. V. (2005). The public and nanotechnology: How citizens make sense of emerging technologies. *Journal of Nanoparticle Research*, *7*, 659-667. DOI: 10.1007/s11051-005-7526-2.

Schutz, H. & Wiedeman, P. M. (2008). Framing effects on risk perception of nanotechnology. *Public Understanding of Science*, *17*(3), 369-379. DOI: 10.1177/0963662506071282

Siegrist, M., Cousin, M. E., Kastenholz, H. & Wiek, A. (2007). Public acceptance of nanotechnology foods and food packaging: The influence of affect and trust. *Appetite, 49*(2), 459-466.

TA Swiss – Centre for Technology Assessment. (2006). Public Reactions to Nanotechnology in Switzerland: Report on Publifocus Discussion Forum "Nanotechnology, Health and Environment", http://www.ta- swiss.ch/a/nano_pfna/2006_TAP8_Nanotechnologien_e.pdf. Accessed February 28, 2009.

REFERENCES

Adams, J. (2003). Risk and morality: Three framing devices. In R. V. Ericson & A. Doyle (Eds.), *Risk and Morality.* Toronto, Canada: University of Toronto Press.

Allum, N., Sturgis, P., Tabourazi, D. & Brunton-Smith, I. (2008). Science knowledge and attitudes across cultures: A meta-analysis. *Public Understanding of Science, 17,* 35-54. DOI: 11.1077/0963662506070159

Anderson, A. (2003). Risk, terrorism, and the Internet. *Knowledge Technology Policy, 16*(2), 24-33. DOI: 10.1007/s12130-003-1023-7

Bardossy, A., Bogardi, I & Duckstein, L. (1993). Fuzzy nonlinear regression analysis of dose-response relationships. *European Journal of Operational Research, 78,* 173-183.

Bakir, V. (2006). Policy agenda setting and risk communication: Greenpeace, Shell, and issues of trust. *Harvard International Journal of Press/Politics, 11*(3), 67-88. DOI: 10.1177/1081180X06289213

Berube, D. (2000). Debunking mini-max reasoning. *Contemporary Argumentation and Debate, 21,* 53-73.

Berube, D. (2006). *Nano-Hype: The Truth Behind the Nanotechnology Buzz.* Amherst, NY: Prometheus Books.

Berube, D. (2007). Rhetoric of stakeholding. In F. Allhoff, P. Lin, J. Moor, & J. Weckert (Eds.), *Nanoethics: The Ethical and Social Implications of Nanotechnology.* San Fransisco, CA: John Wiley and Sons.

Berube, D. (2008a). Rhetorical gamesmanship in the nano debates over sunscreens and nanoparticles. *Journal of Nanoparticle Research.* DOI: 10.1007/s11051-008- 9362-7

Berube, D. (2008b). Stakeholder participation in nanotechnology policy debates. In D. Bennett-Woods (Ed.), *Nanotechnology: Ethics and Society.* Boca Raton, FL: CRC Press.

Berube, D. (2008c). Intuitive toxicology: The public perception of nanoscience. In F. Allhoff & F. Lin (Eds.), *Nanotechnology & Society* (91-108). New York, NY: Springer.

Berube, D. (2009). Public acceptance of nanomedicine: A personal perspective.

Nanomedicine and Nanobiotechnology, NY: Wiley Interdisciplinary Reviews, 1, January/ February 2009, 2-5.

Bimber, B. (1996). *The Politics of Expertise in Congress: The Rise and Fall of the Office of Technology Assessment.* Albany, New York: State University of New York Press.

Bishop, B. (2008). *Why the clustering of like-minded America is tearing us apart.* Houghton Mifflin.

Blakemore, C. (2000). The next big thing – the reality of risk. *FortuneCity.* http://www.fortunecity.com/emachines/e11/86/tnbg1.html. Accessed July30, 2008.

Blaney, J., Benoit, W. & Brazeal, L. (2002). Blowout: Firestone's image restoration campaign. *Public Relations Review*, *28*(4), 379-392. DOI: 10.1016/S0363- 8111(02) 00163-7

Bleeker, S. (1980). Building the public's trust and the bottom-line together. *Public Relations Quarterly*, *25*(4), 5-7.

Bonnafours-Boucher, M. (2005). Some philosophical issues in corporate governance: The role of property in stakeholder theory. *Corporate Governance*, *5*(2), 34-47. DOI: 10.1108/14720700510569229

Bradbury, J. (1989). The policy implications of differing concepts of risk. *Science, Technology, and Human Values*, *14*(4), 380-399. DOI: 10.1177/016224398901400404

Breakwell, G. (2000). Risk communication: Factors affecting impact. *British Medical Bulletin*, *56*(1), 110-120.

Brunsson, N. *The Irrational Organization.* Chichester, UK: Wiley.

Burns, E. (2006). Blogs suffer in poll on preferred news sources. *Clickz.* http://clickz.com/showpage.html?page=3623588. Accessed June 20, 2008.

Cash, D. W., Clark, W. C., Alcock, F., Dickson, N. M., Eckley, N., Guston, D. H., Jäger, J. & Mitchell, R. B. (2003). Knowledge systems for sustainable development. *Proceedings of the National Academy of Sciences*, *100*(14), 8086-8091.

Cho, H., Choi, H. & Kim, Y. (2002). A risk assessment methodology for incorporating uncertainties using fuzzy concepts. *Reliability Engineering and System Safety. 7*, 173-183. DOI: 10.1016/S0951-8320(02)00158-8

Cialdini, R. B. (1987). Compliance principles of compliance professionals: Psychologists of necessity. In M. P. Zanna, J. M. Olson & C. P. Herman (Eds.), *Social influence: The Ontario symposium* (Vol. 5, 165-184). Hillsdale, NJ: Lawrence Erlbaum.

Cialdini, R. B. (1993). *Influence: science and practice* (3rd edn). New York: HarperCollins.

Cobb, R. & Elder, C. (1971). The politics of agenda-building: An alternative perspective for modern democratic theory. *Journal of Politics*, *33*, 892-915.

Cobb, M. & Macoubrie, J. (2004). Public perceptions about nanotechnology: Risks, benefits, and trust. *Journal of Nanoparticle Research*, *6*(4), 395-405. DOI: 10.1007/s11051-004-3394-4

Cobb, M. D. & Hamlett, P. (2008, June). *The first national citizens' technology forum on converging technologies and human enhancement: Adapting the Danish consensus conference in the USA*. Paper presented at the 10th Conference on Public Communication of Science and Technology, Malmo, Sweden.

Cochran, M. E. (2004). "Cynical Politics" versus "Scientific Truth": Communication crisis at Brookhaven National Lab. *IEEE Transactions on Professional Communication*, *47*(4), 256-267.

Cohen, B. (1985). Criteria for technology acceptability. *Risk Analysis*, *5*(1), 1-3. DOI: 10. 1111/j.1539-6924.1985.tb00144.x

Cohen, S. (1972). *Folk Devils and Moral Panics.* London, UK: MacGibon and Kee.

Commission of the European Communities. (2000). White paper on food safety.

COM(1999) 719 final, 12 January 2000. Brussels: Commission of European Communities. http://ec.europa.eu/dgs/health_consumer/library/pub/pub06_en.pdf. Accessed January 04, 2010.

Corbin, R. (1980). Decisions that might not get made. In T. Wallsten (Ed.), *Cognitive Processes in Choice and Decision-making.* Hillsdale, NJ: Elrbaum.

Covello, V., Peters, R., Woitecki, J. & Hyde, R. (2001). Risk communication, the West Nile Virus epidemic, and bioterrorism: Responding to the communication challenges posed by the intentional or unintentional release of a pathogen in an urban setting. *Journal of Urban Health, 78*(2), 383-391. DOI: 10.1093/jurban/78.2.382

Dake, A. (1991). Orienting dispositions in the perception of risk. *Journal of Cross- cultural Psychology, 22*(1), 61-82. DOI: 10.1177/0022022191221006

Darbra, R. & Eljarrat, D. (2008). How to measure uncertainties in environmental risk assessment. *Trends in Analytical Chemistry, 27*(4), 377-385. DOI: 10.1016/j.trac.2008.02.005

deLeon, P. (1990). Participatory policy analysis: Prescriptions and precautions. *Asian Journal of Public Administration, 12*, 29-54.

deLeon, P. (1997). *Democracy and the Policy Science*. Albany: State University of New York Press.

Douglas, M. & Wildavsky, A. (1982). *Risk and Culture.* Berkeley, CA: University of California Press.

Drexler, K.E. (1986) *Engines of Creation: The Coming Era of Nanotechnology*. New York: Anchor Press/Doubleday.

Earle, T. & Cvetkovich, G. (1995). *Social trust: Toward a cosmopolitan society.* Westport, Connecticut: Greenwood Publishing Group.

Eckersley, R. (2001). Losing the battle of the bulge: causes and consequences of increasing obesity. *Medical Journal of Australia..* http://www.mja.com.au/public/issues/174_11_040601/eckersley/eckersley.html. Accessed June 23, 2008.

Edmunds, A. & Morris, A. (2000). The problem of information overload in business organisations: A review of the literature. *International Journal of Information Management, 20*(1),17-28. DOI: 10.1016/S0268-4012(99)00051-1

Elkington, J. & Trisoglio, A. (1996). Developing realistic scenarios for the environment: Lessons from Brent Spar. *Long Range Planning, 29*(6), 762-769.

ETC. (2004). Nano's troubled waters: Latest toxic warning shows nanoparticles cause brain damage in aquatic species and highlights need for moratorium on the release of new nano materials. http://64.233.169.104/search?q=cache:jZI1xpXgG8MJ:www.etcgroup.org/upload/publication/116/01/gt_troubledwater_april1.pdf+2003,+the+ETC+group&hl=en&ct=clnk&cd=1&gl=us .

Etzioni, A. & Diprete, T. (1979). The decline of confidence in America: The prime factor, a research note. *Journal of Applied Behavioral Science, 15*, 520-526.

European Commission. (2001). *Europeans, Science and Technology*. Eurobarometer Special Survey 154, organized and supervised by DG Press and Communication, Brussels.

Evans, W. & Graham, J. (1991). Risk reduction or risk compensation? The case of mandatory safety-belt use laws. *Journal of Risk and Uncertainty, 4*(1), 61-73.

Faber, B. (2007). *Discourse, technology & change*. New York, NY: Continuum.

Financial Times. (2000, July 29/30). Living dangerously. http://pages.stern.nyu.edu/~rramacha/2000-Q3.htm. Accessed December 28, 2004.

Finucane, M., Alhakami, A., Slovic, P. & Johnson, S. (2000). The affect heuristic in judgments of risks and benefits. *Journal of Behavioral Decision-making, 13*, 1-17.

Fischoff, B., Slovic, P., Lichtenstein, S., Reads, S. & Combs, B. (1978). How safe is safe enough? *Policy Sciences, 9*(2), 127-152.

Fisher, E., Mahajan, R. & Mitcham, C. (2006). Midstream modulation of technology: Governance from within. *Bulletin of Science, Technology & Society*, *26*(6), 485- 496. DOI: 10.1177/0270467606295402

Fishkin, J. (1991). *Democracy and Deliberation: New Directions for Democratic Reform.* New Haven, CT: Yale University Press.

Fishkin, J. (1995). *The Voice of the People: Public Opinion and Democracy.* New Haven, CT: Yale University Press.

Fishkin, J. (1996). The televised deliberative poll: An experiment in democracy. *Annals of the American Academy of Political and Social Science*, *546*, 132-140. DOI: 10.1177/ 00027 16296546001012

Fishkin, J. (1997). *The voice of the people – Public opinion and democracy.* New Haven, CT: Yale University Press.

Fitzpatrick, M. (2001). *The tyranny of health: Doctors and the regulation of lifestyle.* New York, NY: Routledge.

Florida, R. (2008). *Who's your city?* New York, NY: Random House.

Flynn, J., Slovic, P. & Kunreuther, H. (Eds.). (2001). *Risk, Media, and Stigma: Understanding Public Challenges to Modern Science and Technology.* London: Earthscan.

Flynn, J., Slovic, P. & Mertz, C. K. (1993). Decidedly different: Expert and public views of risks from a radioactive waste repository. *Risk Analysis*, *13*(6), 643-648.

Freudenberg, W. & Pastor, S. (1992). Public responses to technological risks: Toward a sociological perspective. *Sociological Quarterly*, *33*, 389-412. DOI: 10.1111/j.1533-8525.1992.tb00381.x

Frewer, L., Howard, C., Hedderly, D. & Shepherd, R. (1996). What determines trust in information about food-related risks? Underlying psychological constructs. *Risk Analysis*, *16*(4), 473-486. DOI: 10.1111/j.1539-6924.1996.tb01094.x

Frewer, L., Miles, S., Brennan, M., Kuznesof, S., Ness, M. & Ritson, C. (2002). Public preferences for informed choice under conditions of risk uncertainty. *Public Understanding of Science*, *11*(4), 363-372. DOI: 10.1088/0963-6625/11/4/304

Friedman, S. M., Dunwoody, S. & Rogers, C. L. (1999). *Communicating uncertainty : Media coverage of new and controversial science.* Mahwah, N.J.: Erlbaum Associates.

Galtung, J. & Ruge, M. (1965). The structure of foreign news. *Journal of Peace Research*, 2(1), 64-91. DOI: 10.1177/002234336500200104

Gaskell, G. & Bauer, M. (2006). *Genomics and Society: Legal, Ethical and Social Dimensions.* London: Earthscan.

Gigenrenzer, G. (2002). *Reconing with Risk: Learning to Live with Uncertainty.* London, UK: Allen Line.

Gigerenzer, G. & Goldstein, D. (1996). Reasoning the fast and frugal way: Models of bounded rationality. *Pyschological Review*, *103*, 650-669.

Giles, J. (2005). Internet encyclopedias go head to head. *Nature*, *438*, 900-901. DOI: 10.1038/ 438900a

Goldstein, D. G. & Gigerenzer, G. (2002). Models of ecological rationality: The recognition heuristic. *Psychological Review*, *109*, 75-90.

Goffman, E. (1974). *Frame Analysis: An Essay on the Organization of Experience.* New York, NY: Harper & Row.

Guston, D. (1999). Evaluating the first U.S. consensus conference: the impact of the citizens panel of telecommunications and the future of democracy. *Science, Technology & Human Values*, *24*(4), 451-482.

Guston, D. (2004). Forget politicizing science. Let's democratize science! *Issues in Science & Technology*, *21*(1), 25-28.

Guston, D. & Sarewitz, D. (2002). Real-time technology assessment. *Technology in Society*, *24*, 93-109.

Guston, D., Woodhouse, E. & Sarewitz, D. (2001). *A Science and Technology Policy Focus for the Bush Administration, Issues in Science and Technology*, v. *17*, Spring, 29-32.

Hamlett, P. (2003). Technology theory and deliberative democracy. *Science, Technology & Human Values*, 28(1), 112-140. DOI: 10.1177/0162243902238498 Hamlett, P. & Cobb, M. (2006). Potential solutions to public deliberation problems: structured deliberations and polarized cascades. *The Policy Studies Journal*, *34*(4), 629-648.

Hastie, R. & Park, B. (1986). The relationship between memory and judgment depends on whether the task is memory-based or on-line. *Psychological Review*, *93*, 258- 268. DOI: 10.1037/0033-295X.93.3.258

Hayes, B. & Tariq, V. (2000). Gender differences in scientific knowledge and attitudes toward science: A comparative study of four Anglo-American nations. *Public Understanding of Science*, *9*, 433-447. DOI: 10.1088/0963-6625/9/4/306

Hidi, S. & Baird, W. (1988). Strategies for increasing text-based interest and students' recall of expository texts. *Reading Research Quarterly*, *23*, 465-483.

Ho, M. & Ching, L. (2002). Biotech debacle in four parts. *Institute of Science in Society*. October 24. http://www.i-sis.org.uk/biotechdebacle_updated.php. Accessed June 15, 2008.

Ho, S., Brossard, D. & Scheufele, D. (2008). Effects of value predispositions, mass media use, and knowledge on public attitudes toward embryonic stem cell research. *International Journal on Public Opinion Research*, *20*(2), 171-192. DOI: 10.1093/ijpor/edn017

Huang, C. & Ruan, D. (2008). Fuzzy risks and an updating algorithm with new observations. *Risk Analysis*, *28*(3), 681-694. DOI: 10.1111/j.1539- 6924.2008.01057.x

Hunt, S. & Frewer, L. (2001). Trust in sources of information about genetically modified food risks in the U.K. *British Food Journal*, *103*(1), 46-62.

Hunt, S., Frewer, L. & Shepherd, R. (1999). Public trust in sources of information about radiation risks in the U.K. *Journal of Risk Research*, *2*(2), 167-180. DOI: 10.1080/136698799376916

Iles, A. (2007). Identifying environmental health risks in consumer products: Non-governmental organizations and civic epistemologies. *Public Understanding of Science*, *16*(4), 371-391. DOI: 10.1177/0963662505059442

Iyengar, S. & Kinder, D. (1987). *News That Matters: Television and American Opinion*. Chicago, IL: University of Chicago Press.

Jasanoff, S. (1990). *The fifth branch: Science advisors as policy makers*. Cambridge, MA: Harvard University Press.

Jasanoff, S. (1993). Bridging the two cultures of risk analysis. *Risk Analysis*, *13*(2), 123- 129.

Jefferson Center. (2008). http://www.jefferson-center.org/index.asp?Type=B_Basic& Sec= % 7B2BD10C3C-90AF-438C-B04F-88682B6393BE%7D. Accessed June 15, 2008.

Jenkins, H. (2006). *Convergence culture: Where old and digital media collide.* New York, NY: University Press.

Jenkins-Smith, H. & Silva, C. (1998). The role of risk perception and technical information in scientific debates over nuclear waste storage. *Reliability Engineering and System Safety, 59*(1), 107-122. DOI: 10.1016/S0951- 8320(97)00131-2

Jensen, K. K., Lassen, J., Robinson, P. & Sandøe, P. (2005). Lay and expert perceptions of zoonotic risks: Understanding conflicting perspectives in the light of moral theory. *International Journal of Food Microbiology, 99*, 245-255.

Jones, K. (1998). The British Conundrum. *Reciprocal Meat Conference Proceedings, 51.*

Kahan, D., Slovic, P., Braman, D. & Gastil, J. (2006). Fear of democracy: A cultural evaluation of Sunstein on risk. *Harvard Law Review, 119*, 1071-1109.

Kahan, D., Slovic, P., Braman, D., Gastil, J. & Cohen, G. (2007). *Affect, values, and nanotechnology risk perceptions: An experimental investigation* (March 7, 2007). GWU Legal Studies Research Paper No. 261, Yale Law School, Public Law Working Paper No. 115; GWU Law School Public Law Research Paper No. 261; 2nd Annual Conference on Empirical Legal Studies Paper. Available at SSRN: http://ssrn.com/abstract=968652

Kahan, D. M., Slovic P., Braman, D., Gastil, J., Cohen, G. & Kysar, D. (2008). *Biased assimilation, polarization, and cultural credibility: An experimental study of nanotechnology risk perceptions*: (February 4, 2008). Harvard Law School Program on Risk Regulation Research Paper No. 08-25. Available at SSRN: http//:ssrn.com/ abstract=1090044.

Kahneman, D. (2003). Maps of bounded rationality: A perspective on intuitive judgment and choice. In *Les Prix Nobel: The Nobel Prizes 2002* (449-489). Stockholm, Sweden: Nobel Foundation.

Kahneman, D. (1999). Objective happiness. In D. Kahneman, E. Diener, and N. Schwartz (Eds.), *Well being: the foundations of hedonic psychology* (3-25). New York: Russel Sage.

Kahneman, D. & Tversky, A. (1973). On the psychology of prediction. *Psychological Review*, 80, 237-251.

Kahneman, D. & Tversky, A. (1982). The simulation heuristic. In P. Kahneman, P. Slovic & A. Tversky (Eds.), *Judgment under uncertainty: Heuristics and biases* (201-208). New York: Cambridge University Press.

Kahneman, D., Slovic, P. & Tversky, A. (1982). *Judgment under uncertainty: Heuristics and biases*. New York: Cambridge University Press.

Kapferer, J. (1989). A mass poisoning rumor in Europe. *Public Opinion Quarterly, 53*, 467-481.

Kapferer, J. (1990). *Rumors: Uses, interpretations, and images.* New Brunswick: Transaction Publishers.

Kartez, J. D. (1989). Rational arguments and irrational audiences: Psychology, planning, and public judgment. *Journal of American Institute of Planners, 55*(4), 445-456.

Kasperson, R., Jhaveri, N. & Kasperson, J. (2001). Stigma, places, and the social amplification of risk: Toward a framework of analysis. In J. Flynn, P. Slovic, & H. Kunreuther (Eds.), *Risk, Media, and Stigma: Understanding Public Challenges to Modern Science and Technology* (9-27). London, UK: Earthscan.

Kasperson, R., Renn, O., Slovic, P., Brown, H., Emel, J., Goble, R., Kasperson, J. & Ratick, S. (1988). The social amplification of risk: A conceptual framework. *Risk Analysis*, *8*(2),177-187. DOI: 10.1111/j.1539-6924.1988.tb01168.x

Kaufman, G. (1994). Bank contagion: A review of the theory and evidence. *Journal of Financial Services Research*, *8*, 123-150.

Kearney, A. (1994). Understanding global change: A cognitive perspective on communicating through stories. *Climate Change*, *27*, 419-441. DOI: 10.1007/BF01096270

Kinsella, W. (2004). Public expertise: A foundation for citizen participation in energy and environmental decisions. In S. P. Depoe, J. W. Delicath, & M. A. Elsenbeer (Eds.), *Communication and public participation in environmental decision making* (83-95). Albany, NY: SUNY Press.

Kohn, L., Corrigan, J. & Donaldson, M. (2000). *To err is human: Building a safer health system.* Washington, D.C.: National Academy Press.

Kraus, N., Malmfors, T. & Slovic, P. (1992). Intuitive toxicology: Expert and lay judgments of chemical risks. *Risk Analysis*, *12*, 215-231. DOI: 10.1111/j.1539- 6924.1992.tb00669.x

Krimsky, S. (2007). Risk communication in the Internet age: The rise of disorganized skepticism. *Environmental Hazards*, *7*(2), 157-164. DOI: 10.1016/j.envhaz.2007.05.006

Kruger, J., Wirtz, D., Van Boven, L. & Altermatt, T. W. (2004). The effort heuristic. *Journal of Experimental Social Psychology*, *40*, 91-98.

Kuzma, J., Romanchek, J. & Kokotovich, A. (2008). Upstream oversight assessment for agrifood nanotechnology: A case studies approach. *Risk Analysis*, *28*(4) 1081-1098. DOI: 10.111/j.1539-6924.2008.01071.x

Lambro, D. (1980). *Fat City: How Washington Wastes Your Taxes.* Southbend, IN: Regnery Gateway, Inc.

Langfield, A. (2002). Democratizing journalism. *Online Journalism Review* .

Lazarsfeld, P., Berelson, B. & Gaudet, H. (1944). *The people's choice: How the voter makes up his mind in a presidential campaign.* New York: Columbia University Press.

Lee, C. J., Scheufele, D. A. & Lewenstein, B. V. (2005). Public attitudes toward emerging technologies – Examining interactive effects of cognitions and affect on public attitudes toward nanotechnology. *Science Communication*, *27*(2), 240-267. DOI: 10.1177/1075547005281474

Leiss, W. & Powell, D. (1997). *Mad Cows and Mother's Milk.* 2nd ed. Montreal: McGill University Press.

Lewis, D. (1996). *Dying for Information?* London: Reuters Business Information.

Löfstedt, R. & Renn, O. (1997). The Brent Spar controversy: An example of risk communication gone wrong. *Risk Anaysis*, *17*(2), 131-136.

Luskin, R., Fishkin, J. & Jowell, R. (2002). Considered opinions: deliberative polling in Britain. *British Journal of Political Science*, *32*, 455-487. DOI: 10.1017/S000712340 2000194

Lyn, T. E. (2009). Tiny particles linked to permanent long damage. *Shanghai Daily,com.* http://www.shanghaidaily.com/sp/article/2009/200908/20090820/article_411236.htm. Accessed August 20, 2009.

MacGregor, D., Slovic, P. & Malmfors, T. (1999). How exposed is exposed enough? Lay inferences about chemical exposure. *Risk Analysis*, *19*, 649-659. DOI: 10.1023/A:1007033619806

MacNaughten, P., Kearnes, M. & Wynne, B. (2005). Nanotechnology, governance, and public deliberation: What role for the social sciences? *Science Communication*, *27*(2), 268-291. DOI: 10.1177/1075547005281531

Masuda, J. & Garvin, T. (2006). Place, culture, and the social amplification of risk. *Risk Analysis*, *26*(2), 437-454. DOI: 10.1111/j.1539-6924.2006.00749.x

Mayer, I. (1997). *Debating Technologies: A Methodological Contribution to the Design and Evaluation of Participatory Policy Analysis.* Tilburg, Netherlands: Tilburg University Press.

Maynard, A. (2008). Living with nanoparticles. *Nano Today*, *3*, 1-2.

Mazur, A. (1990). Nuclear power, chemical hazards, and the quantity of reporting. *Minerva*, *28*(3), 294-323. DOI: 10.1007/BF01096293

McCaskey, M. (1982). *The Executive Challenge: Managing Change and Ambiguity.* Boston, MA: Harper Business.

McGrew, T. (2003). Confirmation, heuristics, and explanatory reasoning, *British Journal for the Philosophy of Science*, *54*, 553-567.

McLeod, J. M., Scheufele, D. A., Moy, P., Horowitz, E. M., Holbert, R. L., Zhang, W. W., et al. (1999). Understanding deliberation—The effects of discussion networks on participation in a public forum. *Communication Research*, *26*(6), 743-774

McQuail, D. (2005). *Mass Communication Theory* (5th Edition ed.). London, UK: Sage. Mertens, J. (2006). Smoking gun: When real-life CSI and Hollywood collide. *Law Enforcement Technology*, *33*(3), 52, 54-56, 58, and 61.

Mooney, C. (2005). *The Republican War on Science*. New York, NY: Basic Books. Morrissey, S. (2004). Paving the way for nanotech. *Chemical and Engineering News*, *82*(24), 17.

Nano Risk Framework. (2007). http://edf.org/documents/6496_Nano%20Risk%20Framework.pdf. Accessed June 13, 2008.

Nano Jury UK. (2005). Our Provisional Recommendations. September. http://www.nanojury.org.uk/pdfs/recommendations.pdf. Accessed February 28, 2009.

National Science and Technology Council. (2007). The National Nanotechnology Initiative: Strategic Plan, December 2007. Washington DC: Nanoscale Science, Engineering, and Technology Subcommittee. http://www.nano.gov/NNI_Strategic_Plan_2007.pdf. Accessed January, 4, 2010.

Nehnevajsa, J. (1984). Low-probability/high-consequence risks: Issues in credibility and acceptance. In R. Waller and V. Covello (Eds.), *Low Probability High Consequence Risk Analysis: Issues, Methods, and Case Studies* (521-529). New York, NY: Plenum Press.

New BlogHer study shows U.S. women increasingly shifting to blogs as a mainstream media and communication channel. (2008). *PR-Inside.com.* http://www.pr- inside.com/new-blogher-study-shows-u-s-women-r536690.htm. Accessed June 14, 2008.

Newton, K. (1999). Mass media effects: Mobilization or media malaise? *British Journal of Political Science*, *29*(4), 577-599. DOI: 10.1017/S0007123499000289

Nisbet, M. & Scheufele, D. (2007). The future of public engagement. *The Scientist*, *21*(10), 38-44.

Norris, P. (1996). Does television erode social capital? A reply to Putnam. *PS: Political Science and Politics*, *29*(3), 474-480.

Norris, P. (2000). *A virtuous circle: Political communications in postindustrial societies.* New York, NY: Cambridge University Press.

Norris, P. (2001). *Digital Divide: Civic Engagement, Information Poverty, and the Internet Worldwide.* New York, NY: Cambridge University Press.

Oberdörster, G., Oberdörster, E. & Oberdörster, J. (2005). Nanotoxicology: An emerging discipline evolving from studies of ultrafine particles. *Environmental Health Perspectives, 113*, 823-831.

Orasanu, J. & Connolly, T. (1993). The re-invention of decision-making. In G. Klein, J. Orasanu, R. Calderwood and C. Zambok (Eds.), *Decision-Making in Action: Models and Methods* (3-20). Norwood, NJ: Albex Publishing.

O'Neill, B. & Williams, A. (1998). Risk homeostasis hypothesis: A Rebuttal. *Injury Prevention. 4*. 92-93.

O'Sullivan, K. (2000). Wall criticizes EU White Paper on food safety plans. *Irish Times*. http://www.icel.ie/press_four.htm. Accessed December 28, 2004.

Overbye, D. (2008). Gauging a collider's odds of creating a black hole. *New York Times*. April 15. http://www.nytimes.com/2008/04/15/science/15risk.html. Accessed September 12, 2008.

Peltzman, S. (1975). The effects of automobile safety regulation. *The Journal of Political Economy, 83*(4), 677-726. DOI: 10.1086/260352

Pennington, N. & Hastie, R. (1986). Evidence evaluation in complex decision-making. *Journal of Personality and Social Psychology, 51*, 242-258.

Pennington, N. & Hastie, R. (1988). Explanation-based decision-making: The effects of memory structure on judgment. *Journal of Experimental Psychology: Learning, Memory, and Cognition, 14*, 521-533.

Peters, R., Covello, V. & McCallum, D. (1997). The determinants of trust and credibility in environmental risk communication: An empirical study. *Risk Analysis, 17*(1), 43-54. DOI:10.1111/j.1539-6924-1997.tb00842.x

Petts, J. & Brooks, C. (2006). Expert conceptualizations of the role of lay knowledge in environmental decision-making: Challenges for deliberative democracy. *Environment and Planning, 38*, 1045-1059.

Petty, R. & Cacioppo, J. (1986). *Communication and Persuasion: Central and Peripheral Routes to Attitude Change.* New York, NY: Springer-Verlag.

Pew Research Center for People and the Press. (2008). *Audience segments in a changing news environment: Key news audiences now blend online and traditional sources.* http://people-press.org/report/444/new-media. Accessed August 17, 2008.

Pfau, M., Kenski, H. C., Nitz, M. & Sorenson, J. (1990). Efficacy of inoculation strategies in promoting resistance to political attack messages: Application to direct mail. *Communication Monographs, 57*, 1-12.

Pidgeon, N., Kasperson, R. & Slovic, P. (2003). *The Social Amplification of Risk.* Cambridge, UK: Cambridge University Press.

Poland, C. A., Duffin, R., Kinloch, I., Maynard, A., Wallace, W. A. H., Seaton, A., Stone, V., Brown, S., MacNee, W. & Donaldson, K. (2008). Carbon nanotubes introduced into the abdominal cavity of mice show asbestos-like pathogenicity in a pilot study. *Nature Nanotechnology, 3*, 423-428. doi:10.1038/nnano.2008.111

Poortinga, W. & Pidgeon, N. (2003). Exploring the dimensionality of trust in risk regulation. *Risk Analysis, 23*(5), 961-972.

Poortinga, W. & Pidgeon, N. (2005). Trust in risk regulation: Cause or consequence of the acceptability of GM food? *Risk Analysis*, *25*(1), 199-209. DOI: 10.1111/j.0272-4332.2005.00579.x

Poortinga, W. & Pidgeon, N. (2006). Prior attitudes, salient value similarity, and dimensionality: Toward an integrative model of trust in risk regulation. *Journal of Applied Social Psychology*, *36*(7), 1674-1700. DOI: 10.1111/j.0021- 9029.2006.00076.x

Powell, M. & Kleinman, D. (2008). Building citizen capacities for participation in nanotechnology decision-making: the democratic virtues of the consensus conference model. *Public Understanding of Science*, *17*, 329-348. DOI: 10.1177/0963662506068000

Price, P., Pentecost, H. C. & Voth, R. D. (2002). Perceived event frequency and the optimistic bias: Evidence for a two-process model of personal risk judgments. *Journal of Experimental Social Psychology*, *38*, 242-252. DOI: 10.1006/jesp.2001.1509

Price, V. & Czilli, E. (1996). Modelling patterns of news recognition and recall. *Journal of Communication*, *46*, 53-78. DOI: 10.1111/j.1460-2466.1996.tb01474.x

Purvis-Roberts, K. L., Werner, C. A. & Frank, I. (2007). Perceived risks from radiation and nuclear testing near Semipalatinsk, Kazakhstan: A comparison between physicians, scientists, and the public. *Risk Analysis*, *27*(2), 291-302.

Putnam, R. (1995). Bowling alone: America's declining social capital. *Journal of Democracy*, *6*, 65-78.

Putnam, R. (2000). *Bowling alone: The collapse and revival of American community.* New York: Simon & Schuster.

Rejeski, D. (2005). Introduction. In M. Rodemeyer, D. Sarewicz & J. Wilsdon (Eds.), *The Future of Technology Assessment*, Washington, DC: Woodrow International Center for Scholars. http://www.wilsoncenter.org/news/docs/techassessment.pdf Accessed July 1, 2008.

Renn, O. (1991). Risk communication and the social amplification of risk. In R. Kasperson, & P. Stallen (Eds.), *Communicating risks to the public: International perspectives* (287-324). Dordrecht: Kluwer Academic.

Renn, O. (1992). Concepts of risk: A classification. In S. Krimsky, & D. Golding (Eds.), *Social Theories of Risk* (53-79). Westport, CT: Praeger.

Renn, O. (2003). Hormesis and risk communication. *Human & Experimental Toxicology*, *22*, 3-24. DOI: 10.1191/0960327103ht314oa

Rimmon-Kenan, S. (1983). *Narrative fiction: Contemporary Poetics*. London: Methuen. Rip, A., Misa, T. & Schot, J. (Eds.). (1995). *Managing technology in society*. Pinter: London, England.

Rodemeyer, M., Sarewitz, D. & Wilson, J. (2005). *The Future of Technology Assessment.* Woodrow Wilson International Center for Scholars.

Rodgers, J. (1999). Introduction of section: Overarching considerations in risk communications: Romancing the message. *Journal of the National Cancer Institute Monographs*, *25*, 21-22.

Rogers, R.W. & Mewborn, C.R. (1976). Fear appeals and attitude change: Effects of a threat's noxiousness, probability of occurrence, and the efficacy of coping responses. *Journal of Personality and Social Psychology*, *34(1)*, *54-61.*

Rosati, S. & Saba, A. (2004). The perception of risks associated with food-related hazards and the perceived reliability of sources of information. *International Journal of Food Science Technology*, *39*(5), 491-500. DOI: 10.1111/j.1365- 2621.2004.00808.x

Rothman, S. & Lichter, S. (1987). Elite ideology and risk perception in nuclear energy policy. *The American Political Science Review*, *81*(2), 383-404.

Rowe, G. & Wright, G. (2001). Differences in expert and lay judgment of risk: myth or reality? *Risk Analysis*, *21*(2), 341-356.

Samuels, R. & Stich, S. (2004). Rationality & psychology. In A. R. Mele & P. Rawling (Eds.), *The Oxford Handbook of Rationality* (279-300). Oxford, UK: Oxford University Press.

Sandman, P (1993). *Responding to community outrage: Strategies for effective risk communication.* Fairfax, VA: American Industrial Hygiene Association.

Satterfield, T. Slovic, P. & Gregory, R. (2000). Narrative valuation in a policy judgment context. *Ecological Economics*, *34*, 315-331. DOI:10.1016/S0921-8009(00)00177-4

Savadori, L., Rumiati, R. & Bonini, N. (1998). Expertise and regional differences in risk perception: The case of Italy. *Swiss Journal of Psychology*, *57*(2), 101-113.

Savadori, L., Savio S., Nicotra E., Rumiati R., Finucane M. & Slovic P. (2004). Expert and public perception of risk from biotechnology. *Risk Analysis*, *24*, 1289-1299. DOI: 10. 1111/j.0272-4332.2004.00526.x

Schell, J. (1982). *The Fate of the Earth.* New York, NY: Alfred A. Knopf.

Scherer, C. W. & Cho, H. (2003). A social network contagion theory of risk perception. *Risk Analysis*, *23*(2), 261-267.

Scheufele, D. (1999). Framing as a theory of media effects. *Journal of Communication*, *49* (1), 103-122. DOI: 10.1111/j.1460-2466.1999.tb02784.x

Scheufele, D. (2000). Agenda-setting, priming, and framing revisited: Another look at cognitive effects of political communication. *Mass Communication & Society*, *3* (2&3), 297-316. DOI: 10.1207/S15327825MCS0323_07

Scheufele, D. (2006). Messages and heuristics: How audiences form attitudes about emerging technologies. In J. Turney (Ed.), *Engaging Science: Thoughts Deeds, Analysis, and Action* (20-25). London, UK: The Wellcome Trust.

Scheufele, D. & Brossard, D. (2008). Nanotechnology as a moral issue? Religion and science. *U.S. AAAS Professional Ethics Report*, *21*(1), 1-3.

Scheufele, D. & Shah, D. (2000). Personality strength and social capitol: The role of dispositional and informational variables in the production of civic participation. *Communication Research*, *27*, 107-131.

Scheufele, D. & Tewksbury, D. (2007). Framing, agenda setting, and priming: The evolution of three media effects models. *Journal of Communication*, *79*(2), 427- 444. DOI: 10.1111/j.1460-2466.2006.00326.x

Scheufele, D. Shanahan, J. & Kim, S. (2002). Who cares about local politics? Media influences on local political involvement, issue awareness, and attitude strength. *Journalism & Mass Communication Quarterly*, *79*(2), 427-444.

Schmidt, C. (2007). "We're Sorry": The healing power of apology—and how two little words can make medicine safer. *Harvard Public Health Review*. http://www.hsph.harvard.edu/ review/fall07/fall07apology.html. Accessed July 15, 2008.

Scott, R. (1992). *Organizations: Rational, natural, and open systems*. (3rd ed.). Englewood Cliffs, NJ: Prentice Hall. Shedden, D. (2008). *New Media Bibliography*: http://www.poynter.org/content/content_view.asp?id=899. Accessed July 14, 2008.

Shibutani, T. (1996). *Improvised news: A sociological study of rumor.* Indianapolis, IN: Bobbs-Merrill.

Shoemaker, P. & Reese, S. (1996). *Mediating the Message: Theories of Influences on Mass Media Content* (2nd Edition ed.). White Plains, NY: Longman.

Short, J. (1984). The social fabric of risk: Toward the social transformation of risk analysis. *American Sociological Review, 49*(6), 711-725.

Short, J. & Rosa, E. (2004). Some principles for siting controversy decisions: Lessons from the U.S. experience with high level nuclear waste. *Journal of Risk Research, 7*(2), 135-152. DOI: 10.1080/1366987042000171276 Shirky, C. (2008). *Here Comes Everybody: The Power of Organizing Without Organizations*. New York: Penguin Press.

Siegrist, M. (2008). Factors influencing public acceptance of innovative food technologies and products. *Trends in Food Science & Technology*, 19(11), 603- 608. DOI: 10.1016/j.tifs.2008.01.017.

Siegrist, M., Keller, C., Kastenholz, H., Frey, S. & Wiek, A. (2007). Laypeople's and expert's perceptions of nanotechnology hazards. *Risk Analysis, 27*, 59-69. DOI: 10.1111/j.1539-6924.2006.00859.x

Sjöberg, L. (1999). Perceived competence and motivation in industry and government. In G. Cvetkovich & R. Löfstedt (Eds.), *Social Trust and the Management of Risk* (89-99). London, UK: Earthscan.

Sjöberg, L. (2001). Limits of knowledge and the limited importance of trust. *Risk Analysis, 21*(1), 189-198. DOI: 10.1111/0272-4332.211101

Sjöberg, L. (2001b). Political decisions and public risk perception. *Reliability Engineering and System Safety, 72*, 115-123. DOI:10.1016/S0951- 8320(01)00012-6

Sjöberg, L. (2004). Local acceptance of a high-level nuclear waste repository. *Risk Analysis, 24*(3), 737-749.

Sjöberg, L. (2006). Myths of the psychometric paradigm and how they misinform risk communication.SSE/EFI Working Paper Series in Business Administration, No. 2006: 10.

Sjöberg, L. (2008). Antagonism, trust, and perceived risk. *Risk Management: An International Journal, 10*(1), 32-55.

Sjöberg, L. & Drottz- Sjöberg, B. (1997). Physical and managed risk of nuclear waste. *Risk: Health, Safety & Environment, 8*(2), 115-122.

Sjöberg, L. & Herber, M. (2008). Too much trust in (social) trust? The importance of epistemic concerns and perceived antagonism. *International Journal of Global Environmental Issues, 8*(1-2), 30-44. DOI: 10.1504/IJGENVI.2008.017258

Slovic, P. (1986). Informing and educating the public about risk. *Risk Analysis*, 6, 403- 415. DOI:10.1111/j.1539-6924.1986.tb00953.x

Slovic, P. (1987). Perception of risk. *Science, 236*, 280-285. DOI: 10.1126/science.3563507

Slovic, P. (1992). Perception of risk: Reflections on the psychometric paradigm. In S. Krimsky and D. Golding (Eds.), *Social Theories of Risk.* Westport, CT: Praeger.

Slovic, P. (1993). Perceived risk, trust, and democracy. Risk Analysis, 13(6), 675-682. DOI: 10.1111/j.1539-6924.1993.tb01329.x

Slovic, P. (1998). If hormesis exists...implications for risk perception and communication. *Human & Experimental Toxicology, 17*, 439-440.

Slovic, P. (1999). Trust, emotions, sex, politics, and science: Surveying the risk- assessment battlefield. *Risk Analysis, 19*, 689-701. DOI: 10.1111/j.1539- 6924.1999.tb00439.x

Slovic, P., Malmfors, T., Krewski, D., Mertz, C. K., Neil, N. & Bartlett, S. (1995). Intuitive toxicology. II. Expert and lay judgments of chemical risks in Canada. *Risk Analysis, 15*(6), 661-675.

Slovic, P. & Peters, E. (1998). The importance of worldviews in risk perception. *Journal of Risk Decision and Policy, 3*, 165-170.

Slovic, P., Fischoff, B. & Lichtenstein, S. (1980). Facts and fears: Understanding perceived risks. In R. Schwing & W. Albers Jr. (Eds.), *Societal Risk Assessment: How Safe is Safe Enough?* (181-216). New York, NY: Plenum.

Smith, D. (1990). Beyond contingency planning: Towards a model of crisis management. *Industrial Crisis Quarterly, 4*, 263-275.

Smithson, M. (1989). *Ignorance and Uncertainty: Emerging Paradigms*. New York: Springer-Verlag.

Smithson, M. & Verkuilen, J. (2006). Fuzzy set theory – Applications in the social sciences. *Series: Quantitative Applications in the Social Sciences*, Volume *147*. Thousand Oaks, CA: Sage Publications.

Song, Y., Li, X. & Du, X. (2009). Exposure to nanoparticles is related to pleural effusion, pulmonary fibrosis, and granuloma. *European Respiratory Journal, 34*, 559-567.

Speers, T. & Lewis, J. (2004). Journalists and jabs: Media coverage of the MMR vaccine. *Communication Medicine, 1*(2), 171-181.

Sturgis, P., Roberts, C. & Allum, A. (2005). A different take on the deliberative poll: Information, deliberation, and attitude. *Public Opinion Quarterly, 69*(1), 30-65. DOI: 10.1093/poq/nfi005

Sunstein, C. (2005). *Laws of Fear: Beyond the Precautionary Principle.* Cambridge, UK: Cambridge University Press.

Thompson, J. (1967). *Organizations in action.* New York, NY: McGraw-Hill.

Top Sites United States (2007). *Alexa: The Web Information Company.* http://www.alexa.com/site/ds/top_sites?cc=US&ts_mode=country&lang=none. Accessed July 14, 2008.

Tyler, T. (2006). The state of trust today. *Forbes*: http://www.forbes.com/2006/09/25/trust-government- institutionstech_cx_tt_06trust_0925state.html . Accessed July 21, 2008.

Tversky, A. & Kahneman, D. (1974). Judgment under uncertainty: heuristics and biases. *Science, 185*, 1124-1131.

Uslaner, E. (2004). Trust, civic engagement, and the Internet. *Political Communication, 21*(2), 223-242. DOI: 10.1080/10584600490443895

Viscusi, W. K. (1997). Alarmist decisions with divergent risk information. *The Economic Journal, 107*(445): 1657-1670.

Wahlberg, A. & Sjöberg, L. (2000). Risk perception and the media. *Journal of Risk Research, 3*(1), 31-50. DOI: 10.1080/136698700376699

Waller, R. & Covello, V. (1984). *Low-probability /high-consequence risk analysis.* New York: Plenum.

WASH-1400. (1975). Reactor Safety Study—An Assessment of Accident Risks in U.S. Commercial Nuclear Power Plants. NUREG-75/014, Washington, D.C.: Nuclear Regulatory Commission.

Weaver, R. (1953). *The Ethics of Rhetoric.* Chicago, IL: Henry Regnery.

Weingart, P., Engels, A. & Pansegrau, P. (2000). Risks of communication: Discourses on climate change in science, politics, and the mass media. *Public Understanding of Science, 9*(3), 261-283.

Weinstein, N. (1988). *Attitudes of the Public and the Department of Environmental Protection Toward Environmental Hazard, Final Report.* New Jersey Department of Environmental Protection.

Welch, E. & Hinnant, C. (2003). Internet use, transparency, and interactivity effects on trust in government. *Proceedings of the 36th Hawaii International Conference on System Sciences*, (144-151).

Weiss, R. (2006). Nanotech product recalled in Germany. *Washington Post.* April 6, 2006. A02.

White, D. *(1950).* The 'gatekeeper': A case study in the selection of news. *Journalism Quarterly, 27(3), 383-390.*

White, M. & Eiser, J. (2005). Information specificity and hazard risk potential as moderators of trust asymmetry. *Risk Analysis*, *25*(5), 1187-1198. DOI: 10.1111/j.1539-6924.2005.00659.x

Wiedemann, A., Grutsch, M. & Schutz, H. (2006). The impacts of precautionary measures and the disclosure of scientific uncertainty on EMF risk perception and trust. *Journal of Risk Research*, *9*(4), 361-372. DOI: 10.1080/13669870600802111

Wilde, G. J. S. (1982). The theory of risk homeostasis: implications for safety and health. *Risk Analysis*, *2.* 209-25.

Wilde, G. J. S. (2001) *Target Risk 2.* Toronto, Canada: PDE Publications.

Wilsdon, J. & Willis, R. (2004). *See-through Science: Why Public Engagement Needs to Move Upstream.* London, UK: Demos.

World Medical Association. (1975). Declaration of Helsinki – Ethical principles for medical research involving human subjects. http://www.wma.net/e/policy/b3.htm

Wright, G., Pearman, A. & Yardley, K. (2000). Risk Perception in the U.K. Oil and Gas Production Industry: Are expert loss-prevention managers‘ perceptions different from those of members of the public? *Risk Analysis*, *20*(5), 681-690.

Wurman, R. (1990). *Information Anxiety: What to Do When Information Doesn't Tell You What You Need to Do.* New York: Bantam.

Wynne, B. (1991). Knowledge in context. *Science, Technology, and Human Values*, *16*, 111-121.

Wynne, B. (1992). Public understanding of science research: New horizons or hall of mirrors? *Public Understanding of Science*, *1*, 37.

Xerox joins IORG to combat information overload. (2008). *Business Wire*, July, 20, 2008.

Yaari, M. E. (1987). The dual theory of choice under risk, *Econometrica*, *55*(1), 95-115.

Zaller, J. & Feldman, S. (1992). A simple theory of survey response: Answering questions versus revealing preferences. *American Journal of Political Science*, *36*(3), 579-616.

End Notes

[1] http://www.psandman.com/articles/holing.htm

[2] http://www.decisionresearch.org/pdf/jlist1997-now.pdf.

[3] http://www.dynam-it.com/lennart/.

[4] http://64.233.169.104/search?q=cache:jZI1xpXgG8MJ:www.etcgroup.org/upload/publication/116/01/gt_troubledwater_april1.pdf+2003,+the+ETC+group&hl=en&ct=clnk&cd=1&gl=us

[5] http://www.youtube.com/watch?v=WiVFR2c1T40.

[6] http://communication.chass.ncsu.edu/nirt/videopresentations.html
[7] http://communication.chass.ncsu.edu/nirt/Deliverables.html.
[8] http://www4.ncsu.edu/~pwhmds/index.html
[9] http://www.nhtsa.dot.gov/.
[10] http://www.osha.gov/.
[11] http://www.epa.gov/.
[12] http://www.cpsc.gov/.

In: Contemporary Risk Communication
Editor: Jon B. Madsen
ISBN: 978-1-62808-464-1

Chapter 2

NANOTECHNOLOGY AND ENVIRONMENTAL, HEALTH, AND SAFETY: ISSUES FOR CONSIDERATION*

John F. Sargent Jr.

SUMMARY

Nanotechnology—a term encompassing nanoscale science, engineering, and technology—is focused on understanding, controlling, and exploiting the unique properties of matter that can emerge at scales of one to 100 nanometers. A key issue before Congress regarding nanotechnology is how best to protect human health, safety, and the environment as nanoscale materials and products are researched, developed, manufactured, used, and discarded. While the rapidly emerging field of nanotechnology is believed by many to offer significant economic and societal benefits, some research results have raised concerns about the potential adverse environmental, health, and safety (EHS) implications of nanoscale materials.

Some have described nanotechnology as a two-edged sword. On the one hand, some are concerned that nanoscale particles may enter and accumulate in vital organs, such as the lungs and brains, potentially causing harm or death to humans and animals, and that the diffusion of nanoscale particles in the environment might harm ecosystems. On the other hand, some believe that nanotechnology has the potential to deliver important EHS benefits such as reducing energy consumption, pollution, and greenhouse gas emissions; remediating environmental damage; curing, managing, or preventing diseases; and offering new safety-enhancing materials that are stronger, self-repairing, and able to adapt to provide protection.

Stakeholders generally agree that concerns about potential detrimental effects of nanoscale materials and devices—both real and perceived—must be addressed to protect and improve human health, safety, and the environment; enable accurate and efficient risk assessment, risk management, and cost-benefit trade-offs; foster innovation and public confidence; and ensure that society can enjoy the widespread economic and societal benefits that nanotechnology may offer. Congressionally-mandated reviews of the National Nanotechnology Initiative (NNI) by the National Research Council and the

* This is an edited, reformatted and augmented version of Congressional Research Services publication RL34614, dated January 20, 2011.

President's Council of Advisors on Science and Technology have concluded that additional research is required to make a rigorous risk assessment of nanoscale materials.

INTRODUCTION

Nanotechnology—a term encompassing nanoscale science, engineering, and technology is focused on understanding, controlling, and exploiting the unique properties of matter that can emerge at scales of one to 100 nanometers.[1] These properties are believed by many to offer substantial economic and societal benefits.

A key issue before Congress regarding nanotechnology is how best to protect human health, safety, and the environment as nanoscale materials and products are researched, developed, manufactured, used, and discarded. While the rapidly emerging field of nanotechnology is believed by many to offer significant economic and societal benefits, some research results have raised concerns about the potential environmental, health, and safety (EHS) implications of nanoscale materials. Potential tools the Federal government might use to address these issues include research and development, regulation, and international engagement.

Some of the properties of nanoscale materials (e.g., small size, high surface area-to-volume ratio) that have given rise to great hopes for beneficial applications have also given rise to concerns about their potential adverse implications for the environment, and human health and safety.[2] With more than 1,000 nanotechnology products reportedly commercially available,[3] there is great interest in protecting the health and safety of the scientists working with nanoscale materials, workers who manufacture the products, consumers who use the products, and members of the general public who may be exposed to nanoparticles, as well as in understanding the environmental impact of nanomanufacturing processes and the use and disposal of nanotechnology products.

Nanoscale particles can result from a variety of different processes. While nanoscale particles can occur naturally (e.g., some particles produced by forest fires, sea spray, volcanoes) and as an incidental by-product of human activities (e.g., some particles contained in welding fumes, diesel exhaust, industrial effluents, cooking smoke), EHS concerns have focused primarily on nanoscale materials that are intentionally designed and produced, often referred to as engineered nanomaterials.

Issues surrounding the potential EHS implications of nanotechnology emerged with the launch in 2000 of the National Nanotechnology Initiative (NNI). The NNI is a multi-agency federal effort to coordinate and expand federal nanotechnology research and development (R&D) efforts. Between FY2001 and FY2011, the federal government has invested approximately $14.2 billion in nanotechnology R&D, including approximately $1.8 billion in FY2011 funded under the current continuing resolution (P.L. 111-322). In addition, by one estimate, U.S. private investment in nanotechnology—$2.7 billion in corporate R&D and $1.0 billion in venture capital investments—exceeded U.S. government funding of $1.9 billion in 2008. Many governments around the world have followed the U.S. lead and established their own national nanotechnology programs. The private sector has invested heavily as well. Global nanotechnology R&D investments—public and private—are estimated to have totaled more than $17 billion in 2008 alone.[4]

Such large investments and intensified efforts to capitalize on these public and private investments have caused some observers (as detailed later in this report) to suggest that there is insufficient information about the potential effects nanotechnology products and manufacturing processes may have on human health, safety, and the environment. They assert a variety of uncertainties, including: how nanoscale particles might be transported in air, water, and soil; how they might react with the environment chemically, biologically, or through other processes; how they might be distributed and deposited; and whether they might accumulate in plants or animals.

Others express the view that concerns about nanotechnology EHS implications are often overgeneralized and overstated. Among the arguments they put forth are that nanoscale materials are frequently embedded in other materials as part of the manufacturing process; that some nanotechnology products, such as semiconductors, have nanoscale features but do not contain nanoscale particles; that nanotechnology materials may replace other materials that have significant and known risks; that some nanoscale particles tend to aggregate or agglomerate in the environment into larger particles that no longer have nanoscale dimensions; and that people are regularly exposed to nanoscale particles produced naturally and as incidental by-products of human activities.

Congressionally-mandated reviews of the NNI by the National Research Council (NRC) and the President's Council of Advisors on Science and Technology (PCAST) have concluded that additional research is required to make a rigorous risk assessment of nanoscale materials. In addition, the NRC warned that, until such information is available, precautionary measures should be taken to protect the health and safety of workers, the public, and the environment.

Nevertheless, most stakeholders agree that these concerns about the potential detrimental effects of nanoscale materials and devices—both real and perceived—must be addressed. Among the issues these stakeholders have identified are characterizing the toxicity of nanoscale materials; developing methods for assessing and managing the risks of these materials; and understanding how these materials move in, and interact with, the environment.

This report identifies the potential environmental, health, and safety opportunities and challenges of nanotechnology; explains the importance of addressing nanotechnology EHS concerns; identifies and discusses nanotechnology EHS issues; and summarizes several options for congressional action, including the nanotechnology EHS-related provisions of selected legislation. The *Appendix* provides an overview of selected federal agencies' roles in the regulation of nanotechnology.

OPPORTUNITIES AND CHALLENGES

Historically, many new technologies have delivered general societal benefits while presenting EHS challenges. For example, automobiles increased personal mobility and provided faster, less expensive transportation of goods, but soon became a leading cause of accidental deaths and injuries, as well as a source of emissions that can damage air quality and may affect the global climate. Similarly, genetically-modified (GM) plants have traits such as greater resistance to pests, pesticides, or cold temperatures that contribute to higher crop yields, while critics argue some GM foods contribute to food allergies and antibiotic resistance.[5]

Like other new technologies, nanotechnology offers potential economic and societal benefits, and presents potential EHS challenges as well. Nanotechnology advocates assert, however, that nanotechnology provides the opportunity to reduce or eliminate known risks by engineering around them. Proponents maintain that nanotechnology also offers the potential for significant EHS benefits, including:

- reducing energy consumption, pollution, and greenhouse gas emissions;
- cleaner, more efficient industrial processes;
- remediating environmental damage;
- curing, managing, or preventing deadly diseases; and
- offering new materials that protect against impacts, self-repair to prevent catastrophic failure, or change in ways that protect or aid soldiers on the battlefield.

For example, nanoscale materials show promise for preventing, detecting, tracking, and removing pollutants. According to the Environmental Protection Agency (EPA):

> nanoscale cerium oxide has been developed to decrease diesel engine emissions; iron nanoparticles can remove contaminants from soil and ground water; and nano-sized sensors hold promise for improved detection and tracking of contaminants.[6]

Potential Nanotechnology Cancer Applications

The NCI Cancer Nanotechnology Plan asserts that nanotechnology can serve as an enabling technology for a variety of cancer-related applications:

- imaging agents and diagnostics that allow clinicians to detect cancer in its earliest, most easily treatable, pre-symptomatic stage;
- systems that provide real-time assessments of therapeutic and surgical efficacy;
- multifunctional, targeted devices capable of bypassing biological barriers to deliver therapeutic agents at high local concentrations directly to cancer cells and tissues that play a critical role in the growth and metastasis of cancer;
- agents capable of monitoring predictive molecular changes and preventing precancerous cells from becoming malignant;
- surveillance systems that detect mutations that may trigger the cancer process and genetic markers that indicate a predisposition for cancer;
- novel methods for managing the symptoms of cancer that adversely impact quality of life; and
- research tools that enable investigators to quickly identify new targets for clinical development and predict drug resistance.

Source: Cancer Nanotechnology Plan: A Strategic Initiative to Transform Clinical Oncology and Basic Research Through the Directed Application of Nanotechnology, National Cancer Institute, National Institutes of Health, Department of Health and Human Services, July 2004.

In the area of human health, scientists assert nanotechnology has the potential for improving disease diagnostics, sensing, monitoring, assessment, and treatment. In particular, the National Cancer Institute (NCI) views nanotechnology as likely to provide revolutionary tools to extend and improve lives. In July 2004, NCI launched a five-year, $145 million initiative focused on applying nanotechnology to the prevention, detection, and treatment of cancer and amelioration of its symptoms. At the initiative's launch, then-NCI Director Andrew von Eschenbach identified nanotechnology as a key component of the agency's strategy for ending death and suffering from cancer by 2015 (see text box, "Potential Nanotechnology Cancer Applications").[7] The NCI has reissued the program for an additional five years and expects to complete an updated plan by the end of 2010.

Some characteristics of nanoscale particles could produce both positive and negative consequences. According to E. Clayton Teague, director of the National Nanotechnology Coordination Office (NNCO),

> the unique properties of these [nanotechnology] materials are a double-edged sword: they can be tailored for beneficial properties, but also have unknown consequences, such as new toxicological and environmental effects.[8]

The following examples illustrate how the same nanotechnology material may be both potentially beneficial and potentially harmful:

- Nanoscale silver is highly effective as an antibacterial agent in wound dressings, clothing, and washing machines, but some have expressed concerns that widespread dispersion of nanoscale silver in the environment could kill microbes that are vital to waste water treatment plants and to ecosystems. Some beneficial bacteria, for example, break down organic matter, remove nitrogen from water, aid in animal digestion, protect against fungal infestations, and even aid some animals in defense against predators.[9]
- Some nanoscale particles may have the potential to penetrate the blood-brain barrier, a structure that protects the brain from harmful substances in the blood but also hinders the delivery of therapeutic agents. The characteristics of certain nanoscale materials may allow pharmaceuticals to be developed to purposefully and beneficially cross this barrier and deliver medicine directly to the brain to treat, for example, a brain tumor.[10] Some critics are concerned, however, that nanoscale particles might unintentionally pass through the blood-brain barrier causing harm to humans and animals.[11]
- Certain nanoscale materials are highly chemically reactive due to their high surface-to-volume ratio.[12] This is a property that might be positively exploited in catalysis, treatment of groundwater contamination, and site remediation. This property also is being explored for use in protective masks and clothing as a defense against chemical and biological agents. However, some research results indicate that the reactivity of some nanoparticles potentially can result in cell damage in animals.[13]
- Carbon nanotubes (CNTs) have potential uses in a wide range of applications (e.g., materials, batteries, memory devices, electronic displays, transparent conductors, sensors, medical imaging). However, some scientists have expressed concerns that some CNTs exhibit properties similar to asbestos fibers and might become lodged in organs (e.g., lungs, kidneys, livers), harming humans and animals.[14]

EHS Concerns About Carbon Nanotubes and Other Fullerenes

Much of the public dialogue about potential risks associated with nanotechnology has focused on carbon nanotubes (CNTs) and other fullerenes (molecules formed entirely of carbon atoms in the form of a hollow sphere, ellipsoid, or tube) since they are currently being manufactured and are among the most promising nanomaterials. These concerns have been amplified by some research on the effects of CNTs on animals and on animal and human cells. For example, researchers have reported that carbon nanotubes inserted into the trachea of mice can cause lung tissue damage;[a] that buckyballs (spherical fullerenes) caused brain damage in fish;[b] and that buckyballs can accumulate within cells and potentially cause DNA damage.[c]

There are scientists who have argued that experiments indicating CNT/fullerene toxicity are not conclusive. They suggest that toxicity reported by researchers may have resulted from uncharacterized contaminants in the samples resulting from the synthesis, purification, and post-processing methods used in the manufacture of CNTs. Thus, they assert, the experiments could be measuring the toxicity of non-nanoscale materials and, therefore, unfairly indicting nanoscale materials. They also contend that such non-nanoscale contaminants, if identified as toxic, potentially could be eliminated or controlled in the manufacturing process. The issue of contaminants is often cited by advocates for improved standards, reference materials, sensors, instrumentation, and other technologies for the characterization of nanoscale materials.

Some experiments have produced results that indicate CNTs/fullerenes are non-toxic. Research on single-walled carbon nanotubes (SWCNTs) by the Institute of Toxicology and Genetics in Karlsruhe, Germany, reported that, in three of four different types of tests conducted, SWCNTs did not show toxicity. In the fourth test, which appeared to indicate SWCNT toxicity, the researchers concluded that the results were a "false positive" and explained how the SWCNTs interacted with the materials in the assay to produce a misleading result. These researchers concluded that this result points to the need for careful selection of assays and the need for the establishment of standards for toxicity testing of CNTs and other nanomaterials.[d]

Work at Rice University's Center for Biological and Environmental Nanotechnology conducted in 2005 found cell toxicity of CNTs to be low, and that it could be reduced further through simple chemical changes to the surface.[e] Earlier research demonstrated that similar surface modifications of buckyballs reduced their toxicity. Nanotechnology may offer the potential to engineer around known and potential hazards by changing the size, molecular construction, or other property of a nanoscale material to make it safe or less hazardous. Experts advise that the potential to do so will require a thorough understanding of the properties of the various nanoparticles and their effects on humans and other organisms.

a Lam, C.W., James, J.T., McCluskey, R., and Hunter, R.L. "Pulmonary toxicity of single-wall carbon nanotubes in mice 7 and 90 days after intratracheal instillation," September 2003. http://www.ncbi.nlm.nih.gov/sites/entrez?cmd= Retrieve&db= Pub Med&list_uids=14514958

b Oberdörster, Eva. "Manufactured Nanomaterials (Fullerenes, C60) Induce Oxidative Stress in the Brain of Juvenile Largemouth Bass," April 2004. http://www.pubmedcentral.nih.gov/articlerender.fcgi?tool=pmcentrez&artid=1247377

c Magrez, A., Kasas, S., Salicio, V., Pasquier, N., Seo, J.W., Celio, M., Catsicas, S., Schwaller, B., and Forro, L. "Cellular Toxicity of Carbon-Based Nanomaterials," *Nano Letters*, 6(6):1121-1125, American Chemical Society, May 2006. http://pubs.acs.org/cgi-bin/abstract.cgi/nalefd/2006/6/i06/abs/nl060162e.html

d Wörle-Knirsch, J.M., Pulskamp, K., and Krug, H. F. "Oops They Did It Again! Carbon Nanotubes Hoax Scientists in Viability Assays," American Chemical Society, *Nano Letters*, Vol.6, April 2006.

e "Modifications render carbon nanotubes nontoxic," press release, Rice University, October 2005.

IMPORTANCE OF ADDRESSING EHS ISSUES

Nanotechnology covers a wide swath of scientific fields, engineering disciplines, and technological applications. Sufficient knowledge has been developed about the useful properties of certain nanomaterials, how they can be manufactured, and how they can be applied in useful ways to enable commercial product development. In other areas of nanotechnology, fundamental research on nanoscale phenomena and processes is under way that may lead to greater understanding and beneficial applications in the years ahead. In general, however, nanotechnology is still an emerging field and there is a dearth of information about how nanoscale particles and devices might adversely affect human health, safety, and the environment. Accordingly, there is widespread agreement on the need for more research to better understand such implications.

In reviews of the NNI,[15] both the 2006 National Research Council and the 2008 President's Council of Advisors on Science and Technology (PCAST) reports concluded that assessment of potential nanotechnology EHS risks was not possible due to the absence of information and tools. According to the NRC,

> it is not yet possible to make a rigorous assessment of the level of risk posed by [engineered nanomaterials]. Further risk assessment protocols have to be developed, and more research is required to enable assessment of potential EHS risks from nanomaterials.[16]

Similarly, PCAST found that

> it is premature to rigorously assess the levels of risk posed by engineered nanomaterials. Adequate tools are being developed but are not yet in place.[17]

Subsequently, in its third assessment of the NNI, the NRC alluded to potential EHS risks, stating:

> Research to date suggests that some products of nanotechnology have the potential to present new or unusual risks to human health and the environment. For instance, nanoscale particles may penetrate to places in the body that are inaccessible to larger particles; radical changes in behavior at the nanoscale may render harmful materials considered to be safe in larger-scale and more conventional forms.[18]

Leaders of the NNI have argued strongly that to achieve the economic, societal, and EHS benefits of nanotechnology the nation must concurrently address its potential adverse effects. According to then-Under Secretary of Commerce for Technology Phillip J. Bond, a leading Bush Administration advocate for the NNI,

> Addressing societal and ethical issues is the right thing to do and the necessary thing to do. It is the right thing to do because as ethically responsible leaders we must ensure that technology advances human well-being and does not detract from it. It is the necessary thing to do because it is essential for speeding technology adoption, broadening the economic and societal benefits, and accelerating and increasing our return on investment.[19]

The NRC's third assessment of the NNI reinforces the perspective that EHS-related uncertainty may stymie nanotechnology innovation and commercialization:

> In the absence of more detailed scientific evidence—and effective assessment and communication of the evidence that does exist—the distinction between plausible and implausible risks remains unclear. The resulting uncertainty threatens to undermine confidence and trust amongst investors, businesses, and consumers, and could jeopardize the success of nanotechnology. This is not a hypothetical threat. Consumer and advocacy groups already have raised concerns over the use of engineered nanomaterials in products as diverse as clothing, fuel additives, and sunscreens. Businesses have been hampered by regulatory uncertainty. A number of industries have shied away from nanotechnology for fear of consumer rejection in the face of speculative concerns.[20]

According to the NRC, the nanotechnology industry and a variety of environmental and public-health interest groups agree that an adequate evaluation of the potential health and environmental effects of engineered nanomaterials is necessary

> to ensure that the future of nanotechnology is not burdened by uncertainties and innuendo
> about potential adverse health and environmental effects of engineered nanoscale materials.[21]

A 2006 survey of business leaders in the field of nanotechnology indicated that nearly two-thirds believed that "the risks to the public, the workforce, and the environment due to exposure to nano particles are 'not known,'" and 97% believed that it is very important or somewhat important for the government to address potential health effects and environmental risks that may be associated with nanotechnology.[22]

The Project on Emerging Nanotechnologies (PEN) has warned that bad practices in nanotechnology research or production may result in a nanotechnology accident that would

> chill investment, galvanize public opposition, and generally lead to a lot of hand wringing on the part of governments who are betting large sums of money on the nanotech revolution.[23]

Successfully addressing EHS issues is seen as vital for those potentially exposed to nanoscale materials (e.g., consumers, researchers, manufacturing workers, the general public), businesses, and investors for a variety of reasons:

- protecting and improving human health, safety, and the environment;
- enabling accurate and efficient risk assessments, risk management, and cost-benefit trade-offs;
- ensuring public confidence in the safety of nanotechnology research, engineering, manufacturing, and use;
- preventing a problem in one application area of nanotechnology from having negative consequences for the use of nanotechnology in unrelated application areas due to public fears, legislative interventions, or an overly-broad regulatory response; and

- ensuring that society can enjoy the widespread economic and societal benefits that nanotechnology is believed by many to offer.

In addition, the U.S. regulatory environment for nanotechnology could be an enabler for innovation and contribute to a strong, sustainable economy by creating predictability, accurately assessing risks and benefits, and fostering the swift movement of safe products into the market. Such an environment is likely to favor nanotechnology-related investments and innovative activities in the United States by domestic and foreign stakeholders, as opposed to nations where such regulatory conditions do not exist.

Conversely, if the U.S. regulatory environment is not handled effectively (i.e., if it lacks predictability, if regulatory approaches do not accurately assess risks and benefits, or if approval processes are too long or expensive) it could prove a major impediment to innovation, economic growth, and job creation, as well as posing a potential threat to health, safety, and the environment. In such a regulatory environment, investment capital may be driven away from nanotechnology, potentially beneficial products may not be developed, safe products may be denied regulatory approval, or unsafe products may be allowed to enter the market.

Alternatively, nanotechnology investments, research, and production may be driven to other nations with preferable regulatory environments. On the one hand, such a regulatory system might be more desirable to investors and companies because it is more predictable, more efficient, and less costly. In such a case, the United States might miss out on nanotechnology's potential economic benefits. On the other hand, if other nations' regulatory systems are more attractive to investors and producers because those systems under-regulate or do not regulate at all, then nanotechnology research, development, and production could present increased EHS risks worldwide.

SELECTED ISSUES FOR CONSIDERATION

Given the widespread agreement that nanotechnology EHS concerns must be addressed, discourse on how best to do so has focused on three main issues:

- federal investment in EHS research;
- federal regulation; and
- international engagement.

These issues are closely interrelated. For example, reliable EHS research is required by regulatory bodies to determine whether and how to regulate nanotechnology products. Since all nations face the same fundamental health, safety, and environmental issues, international coordination on EHS research could help accelerate development of a common body of knowledge through the sharing of results and reduction in redundant research. This shared knowledge could, in turn, inform regulatory decision making and perhaps improve the consistency of regulations among nations. Regulations, standards, and enforcement might need to be coordinated worldwide to protect workers and consumers as intermediate and final products are frequently produced along global supply chains and sold in industrial and commercial markets around the world. In addition, one nation's policies governing nanotechnology production, use, and disposal may have implications for nearby nations and, perhaps, for all nations.

Federal Investment in EHS Research

Current Funding Level

There is not a single, centralized source of EHS research funds that is allocated to individual agencies. Agency nanotechnology budgets are developed internally as part of each agency's overall budget development process. These budgets are subjected to review, revision, and approval by the Office of Management and Budget (OMB) and become part of the President's annual budget submission to Congress. The NNI budget—and the EHS component—is then calculated by aggregating the nanotechnology components of the appropriations provided by Congress to each federal agency. While there is some coordination of EHS-research budget requests through the Nanotechnology Environmental and Health Implications (NEHI) working group[24] and in OMB's budget development process, the decision process that establishes overall funding for nanotechnology EHS research is highly decentralized.

In FY2010, NNI funding for EHS implications research[25] was an estimated $91.6 million, approximately 5.1% of the total NNI budget of $1.781 billion. This represented an increase over the FY2009 EHS research level of $74.5 million[26] (4.4% of the total NNI budget), and the FY2008 level of $67.9 million (4.4%), both in dollars and in share of total NNI funding. President Obama requested $116.9 million (6.6%) for EHS research in FY2011. NNI EHS research funding for FY2006 through FY2010, and President Obama's request for FY2011, is provided in *Table 1.*

Table 1. NNI EHS Research Funding, FY2006-2010, FY2011 Request

	EHS research, in current dollars	EHS research's share of total annual NNI budget
FY2006 (actual)	$ 37.7 million	2.8%
FY2007 (actual)	48.3 million	3.4%
FY2008 (actual)	67.9 million	4.4%
FY2009 regular (actual)	74.5 million	4.4%
FY2009 ARRA (actual)	12.0 million	N/A
FY2010 (actual)	91.6 million	5.1%
FY2011 (request)	116.9 million	6.6%

Source: CRS analysis of data from the FY2008, FY2010, and FY2011 editions of "The National Nanotechnology Initiative: Research and Development Leading to a Revolution in Technology and Industry," NSET Subcommittee, NSTC, The White House, July 2007, and "National Nanotechnology Initiative: FY2009 Budget and Highlights," NSET Subcommittee, NSTC, The White House, February 2008.

NNI officials assert that the initiative also conducts EHS research as a part of its other research activities, but that these EHS investments are not easily quantified and thus are not reflected in the NNI's reported figure for EHS funding. PCAST agreed with this assertion in its 2008 assessment, arguing that

> In many instances, nanotechnology EHS research cannot be separated from the particular application(s) research and from the context for which a specific nanomaterial is intended. Such division is unproductive and neglects the whole benefit of research. Consequently, [PCAST] expects that a substantial fraction of nanotechnology research related to EHS will continue to take place under the auspices of agencies that fund applications R&D and may not be uniquely or exclusively identified as nanotechnology EHS research.... Furthermore, detailed reporting on the degree of relevance to EHS of such research is not necessarily critical to (and may actual hinder) overall prioritization and coordination.[27]

This undercounting was evidenced in part by a one-time OMB request in 2007 to all NNI research agencies to report FY2006 funding data on research related to the five categories identified in the NSET document, *Prioritization of Environmental, Health, and Safety Research Needs for Engineered Nanoscale Materials.*[28] Totals for EHS implications research spending identified in each of the five categories is shown below in **Table 2.** Preliminary analysis of this data by the NEHI working group indicated that NNI agencies spent nearly twice as much on EHS research in FY2006 than was previously reported ($67 million identified by the OMB data-call versus $37.7 million in the President's budget.

Table 2. FY2006 NNI Funding for EHS Research
by Research Needs Categories

Category	Estimated Funding
Instrumentation, Metrology, and Analytical Methods	$27 million
Nanomaterials and Human Health	$24 million
Nanomaterials and the Environment	$13 million
Health and Environmental Exposure Assessment	$ 1 million
Risk Management Methods	$ 3 million
TOTAL	$67 million

Source: Teague, E. Clayton, director, National Nanotechnology Coordination Office. Testimony before the Subcommittee on Research and Science Education, Committee on Science and Technology, U.S. House of Representatives. Hearing on "Research on Environmental and Safety Impacts of Nanotechnology: Current Status of Planning and Implementation under the National Nanotechnology Initiative." 110th Congress, 1st Session, October 31, 2007.
Note: Numbers may not add due to rounding.

Critics (as detailed in the following section) assert that the current level of federal nanotechnology EHS research is too low and represents too small a share of the overall NNI budget. These critics argue that the current allocation of NNI funding may produce a flood of products for which there is inadequate information to assess and manage their EHS risks.

However, executive branch officials stress that the United States leads the world in EHS funding and, by inference, that the current funding level is adequate. White House Office of

Science and Technology Policy (OSTP) director John Marburger asserted that the United States

> leads the world not only in spending for nanotechnology development, but also, by an even larger margin, in its investment in research to understand the potential health and safety issues.[29]

Similarly, NNCO director E. Clayton Teague asserted U.S. leadership in nanotechnology EHS research:

> During fiscal years 2005 through 2008, it is estimated that NNI agencies will have invested nearly $180 million in research whose primary purpose is to address the EHS implications of nanomaterials. With these investments, the United States leads all other countries by a wide margin in support of such research.[30]

In early reviews of the NNI, both the NRC and PCAST concluded that federal EHS research funding should be expanded. According to the NRC assessment,

> To help ensure the responsible development of nanotechnology ... research on the environmental, health, and safety effects of nanotechnology [should] be expanded.[31]

PCAST acknowledged potential EHS risks in its first review of the NNI but found the federal government was "directing appropriate attention" and "adequate resources" to EHS research. In its second assessment, PCAST termed the current federal investment level in EHS "appropriate," but added that

> expanded EHS research, broad-based protocol development, and particularly standardization are necessary.... the funding level for EHS [should] continue to grow consistent with the needs identified in the NNI research strategy for nanotechnology EHS as well as the available capacity for quality research.[32]

Under President Obama, PCAST struck a different tone. In its third assessment of the NNI, PCAST acknowledged the importance of adequate funding and appropriate accounting, but emphasized that

> appropriate and targeted funding for strategic nanotechnology EHS research is more important than absolute dollar amounts. To ensure that emerging EHS issues are addressed effectively and in a way that yields useful information for regulators and policymakers, the NNI needs to help the scientific community establish a substantial core of exploratory research into biological and environmental interactions with nanomaterials. In addition, the Federal Government needs to ensure sufficient funds are available to mission-driven agencies to address specific issues that are arising.[33]

In this regard, PCAST credits the NNI's "substantial funding increases for nanotechnology EHS research" for agencies such as the National Institute for Occupational Safety and Health, Environmental Protection Agency (EPA), Food and Drug Administration (FDA), and Consumer Product Safety Commission (CPSC), noting that:

> Significantly, this will be the first time that FDA and CPSC will have had a specific allocation of funds to cover nanotechnology, a welcome move and one that the NNAP hopes is sustained over a number of years.[34]

Alternative Approaches

Various alternatives have been suggested for addressing the perceived shortcoming in EHS funding. One recommendation is requiring a fixed percentage of the NNI's total funding be devoted to EHS research. A figure of 10% has been proposed for this purpose by organizations such as the NanoBusiness Alliance and the Project on Emerging Nanotechnologies. If this proposal had been in effect in FY2010, the NNI would have been required to spend $178.1 million on EHS research, nearly twice as much as the NSET-reported level of $91.6 million. In testimony before the House Committee on Science and Technology, Sean Murdock, executive director of the NanoBusiness Alliance, agreed with the level of funding represented by the 10% figure but argued the need for cross-agency flexibility in achieving it:

> The NanoBusiness Alliance believes that environmental, health, and safety research should be fully funded and based on a clear, carefully-constructed research strategy. While we believe that 10 percent of the total funding for nanotechnology research and development is a reasonable estimate of the resources that will be required to execute the strategic plan, we also believe that actual resource levels should be driven by the strategic plan as they will vary significantly across agencies.[35]

Others have suggested a different approach, proposing fixed dollar amounts or minimum levels. For example, the Environmental Defense Fund has called for $100 million or more in federal nanotechnology EHS research funding.[36]

In its 2008 assessment, PCAST disagreed with both approaches:

> growing research in nanotechnology EHS must be strategic, guided by ... a comprehensive set of scientifically determined priorities and needs rather than arbitrary percentages or funding figures.[37]

By establishing a 10 percent requirement (or setting a specific dollar figure), the United States could accelerate the growth in EHS research spending. However, in testimony before Congress in 2007, then-PCAST co-chair Floyd Kvamme warned against a rapid increase:

> In general, increasing funding too rapidly does not lead to equivalent increases in high quality research. It is crucial to note that EHS research also depends on advances in non-EHS areas, such as instrumentation development and basic research on nanomaterials.[38]

Some non-governmental organizations (NGOs) have advocated for a more restrained approach to nanotechnology research and development. They assert that the federal government is pushing ahead too quickly in developing nanotechnology and encouraging its commercialization and use without sufficient knowledge and understanding of EHS implications and how they might be mitigated.[39] They argue that the very characteristics that make nanotechnology promising also present significant potential risks to human health and safety and the environment. Some of these groups argue for application of the "precautionary principle,"[40] which holds that regulatory action may be required to control potentially hazardous substances even before a causal link has been established by scientific evidence.[41] In 2006, Friends of the Earth warned that

> The early warning signs surrounding nanotoxicity are serious and warrant a precautionary approach to the commercialization of all products containing nanomaterials ... there should be a moratorium on the further commercial release of sunscreens, cosmetics and personal care products that contain engineered nanomaterials, and the withdrawal of such products currently on the market, until adequate public, peer-reviewed safety studies have been completed, and adequate regulations have been put in place.... [42]

The Action Group on Erosion, Technology, and Concentration (ETC Group) has called for a moratorium on the conduct of nanotechnology R&D and use of commercial products incorporating man-made nanoparticles:

> Given the concerns raised over nanoparticle contamination in living organisms, Heads of State ... should declare an immediate moratorium on commercial production of new nanomaterials and launch a transparent global process for evaluating the socio-economic, health and environmental implications of the technology.[43]

In 2003, the ETC Group expanded the breadth of its proposed moratorium:

> In the absence of toxicology studies, ETC Group believes that governments must also urgently consider extending the moratorium to products that place consumers in direct contact with synthetic nanoparticles through their skin, lungs or digestive systems.[44]

In contrast to these views, a report prepared by the NSET Subcommittee concluded that conducting EHS research in parallel with the development of nanomaterials and their applications will help to ensure the full, safe, and responsible realization of the promise of nanotechnology.[45]

In 2003, then-Under Secretary of Commerce for Technology Phillip J. Bond addressed called for a moratorium or slowdown in nanotechnology R&D, casting the issue in ethical terms:

> Those who would have us stop in our tracks argue that it is the only ethical choice. I disagree. In fact, I believe a halt, or even a slowdown, would be the most unethical of choices.... Given the promise of nanotechnology, how can our attempt to harness its power at the earliest opportunity—to alleviate so many of our earthly ills—be anything other than ethical? Conversely, how can a choice not to attempt to harness its power be anything other than unethical?[46]

Management of Federal EHS Research

Research Priorities and Strategies

In order to manage the Federal EHS portfolio, policymakers will need to establish research priorities. In its first review of the NNI, the NRC recommended that

> Assessing the effects of engineered nanomaterials on public health and the environment requires that the research conducted be well defined and reproducible and that effective methods be developed and applied to (1) estimate the exposure of humans, wildlife, and other ecological receptors to source material; (2) assess effects on human health and ecosystems of both occupational and environmental exposure; and (3) characterize, assess, and manage the risks associated with exposure.[47]

In 2005, PCAST concluded that EHS research should give highest priority to workplace exposure, noting

> the greatest likelihood of exposure to nanomaterials is during manufacture, and therefore [we] agree with the prioritization of research on potential hazards from workplace exposure.[48]

In its 2008 assessment, PCAST reiterated this point stating, "the greatest risk of exposure to nanomaterials at present is to workers who manufacture or handle such material," but also acknowledged a broader range of risks:

> environmental, health, and safety risks in a wide range of settings must be identified and the necessary research performed so that real risks can be appropriately addressed.[49]

In February 2008, the NSET published its much-awaited *Federal Strategy for Environmental, Health, and Safety (EHS) Research Needs for Engineered Nanoscale Materials*. The report describes the NNI's EHS research strategy, identifies lead agencies for each of five research categories, and asserts that it provides "a framework to guide and inform agency efforts to address prioritized research areas and to sustain a diverse program to advance knowledge and support risk decision-making."[50]

Subsequently, the NSET requested the NRC independently review this strategy document. In 2009, the NRC published the results of its review, Review of Federal Strategy for Nanotechnology-related Environmental, Health, and Safety (EHS) Research.[51] While complimentary of the widespread collaboration and coordination required to produce the report and its potential usefulness in "communicating the breadth of federally supported research associated with developing a more comprehensive understanding of the environmental, health, and safety implications of nanotechnology," the NRC review asserted that:

- research needs in risk management and exposure assessment were "poorly defined and incomplete;"
- research needs were not presented as "concrete, measurable objectives" and that no explanation was provided of how success would be measured or the amount of resources required to achieve them;
- the NSET overstates federal funding specifically addressing nanotechnologyrelated EHS issues and that funding may be inadequate;
- the approach used by the NSET for its gap analysis is "flawed and is neither accurate nor complete in laying a foundation for a research strategy"; and
- federal EHS nanotechnology funding is dominated by agencies traditionally focused on exploratory and investigator-driven research (such as NIH and NSF) and that if

these agencies are to continue to lead, their approaches may need to be modified "to ensure that the research they support feeds into an effective EHS risk research strategy based on appropriate, targeted research."

The NRC concluded that what was needed was an effective "national strategy" that involves a range of stakeholders beyond the federal government, including academia, industry, consumer and environmental groups, and others. Such a plan, according to the NRC, would

> identify research needs clearly and estimate the financial and technical resources needed to address identified research gaps. A national strategic plan would be focused on providing solutions to challenges that do not necessarily fit neatly into disciplinary and institutional silos, and ensure important research does not fall between the gaps. Such a plan would also provide specific, measurable objectives and a timeline for meeting them.[52]

Proposals for a Research Roadmap: Differing Perspectives

Some stakeholders assert that a comprehensive approach to federal EHS research has been hampered by the lack of an NNI roadmap for these efforts.[53] In general, these stakeholders seek a multi-year roadmap with specific milestones, metrics, and funding levels. Such a roadmap, they assert, would contribute to a more coordinated approach among agencies and between the executive branch and Congress on the magnitude, timing, prioritization, and management of federal EHS research.

NNI officials argue that the NSET Subcommittee, the coordinating body for the NNI, has developed an EHS research strategy and articulated it in three reports (see text box, "NNI EHS-focused Reports"), though they acknowledge that these documents do not constitute a roadmap. Some Members of Congress have expressed concerns about the time required by the National Nanotechnology Coordination Office to produce a prioritized, detailed implementation plan for NNI EHS research.[54] While acknowledging the challenges faced by the NNCO in developing consensus among the 25 NNI agencies, some Members suggested that these challenges were emblematic of the need for a more top-down approach to EHS research.

NNI EHS-focused Reports

Environmental Health and Safety Research Needs for Engineered Nanoscale Materials, published in September 2006, identified the research and information needed to enable sound risk assessment and risk management decision making with respect to nanoscale materials and products that incorporate them.

Prioritization of Environmental, Health, and Safety Research Needs for Engineered Nanoscale Materials, published in August 2007, identified five broad categories of EHS research and information needs, and five specific research areas in each category.

The National Nanotechnology Initiative: Strategy for Nanotechnology-related Environmental, Health, and Safety Research, published in February 2008, defined the NNI's strategy for addressing priority research on EHS aspects of nanomaterials. The document reviewed current agency research using the taxonomy developed in the second report; identified research gaps; and articulated a framework for prioritizing research, implementing the strategy, and coordinating agency efforts.

Opposition to an EHS roadmap stems primarily from doubts of the practicality and efficacy of such an approach. Some argue that it is unlikely that OMB would commit to a multiyear, multi-agency roadmap accompanied by specific funding levels. Such an approach would depart from the current executive branch annual budget development process and reduce OMB's flexibility in future years. In addition, agencies often have to respond to new requirements based on emergent circumstances, Congressional direction, or other factors. Agency funding is often redirected from planned efforts to new, often imminent, priorities. The need for such redirection of funding could impede the achievement of roadmap milestones and metrics or, conversely, impede the movement of funding to new priorities.

To overcome the obstacles associated with the development of a roadmap by the agencies, some have suggested the National Academies produce such a roadmap. Some experts assert that this approach worked well with respect to the development of a federal research roadmap to reduce EHS uncertainties associated with airborne particulate matter. Others argue that the particulate matter effort focused only a narrow field and covered research conducted by only a single agency (EPA), in contrast, nanotechnology spans a broad range of materials and applications across many fields, and requires EHS research efforts by several agencies.

In February 2007, 19 environmental and business organizations, large and small companies, and research organizations signed a letter to the Senate Appropriations Subcommittee on Interior, Environment, and Related Agencies requesting $1 million be appropriated for the development of a federal roadmap and research strategy. The letter recommended that this work be done by the National Institute of Environmental Health Sciences (NIEHS).[55]

The Senate Appropriations Committee report (S.Rept. 110-91) accompanying the Department of the Interior, Environment, and Related Agencies Appropriations Act, 2008[56] urged the Environmental Protection Agency (EPA) to

> contract or enter into a cooperative agreement with the National Academy of Sciences' Board on Environmental Studies and Toxicology within 90 days of enactment to develop and monitor implementation of a comprehensive, prioritized research roadmap for all Federal agencies on environmental, health and safety issues for nanotechnology.[57]

In July 2009, the National Academies' Board on Environmental Science and Toxicology began an EPA-sponsored project, titled "A Research Strategy for Environmental, Health, and Safety Aspects of Engineered Nanomaterials." According to the National Academies, the project is to produce two reports over four years. The first report, due 18 months from project inception, is to

> present a conceptual framework and priorities for the research program, identify the most important short-term and longer-term research priorities, develop a strategy for monitoring and evaluating research progress, and estimate the resources needed to implement this strategy.[58]

The second report, due at the end of the study period (approximately July 2013), is to

> evaluate research progress and update the research priorities and resource estimates based on results of studies and emerging trends in the nanotechnology industry.[59]

Budget Development, and Coordination and Integration of Efforts

The process used to develop research priorities and the federal EHS budget has also raised management concerns. As discussed earlier, the federal nanotechnology EHS research portfolio results from research funding requests made by individual agencies pursuing their missions and by decisions made in the congressional appropriations process. Informal research coordination among EHS funding agencies occurs through the NEHI working group and more formally through the OMB budget development process.

In its third review of the NNI, PCAST recommended that

> the NSET Subcommittee implement organizational changes that support consequential cross-agency action on addressing nanotechnology EHS issues. In particular, the NNCO should create a senior-level position to lead interagency coordination of efforts in the area of EHS.[60]

In 2010, the NNCO established and filled a new position with the dual titles of Deputy Director and EHS Coordinator. A primary duty of this position is the coordination of EHS research among NNI agencies.

Some proponents for an integrated federal EHS research effort have called for a more top-down approach. The Woodrow Wilson Center's Project on Emerging Nanotechnologies (PEN) has been a leading advocate on this issue. PEN's chief science advisor, Andrew Maynard, asserted that

> to realize nanotechnology's benefits ... the federal government needs a master plan for identifying and reducing potential risks. This plan should include a top-down risk research strategy, dedicated and sufficient funding to do the job, and the mechanisms to ensure that resources are used effectively.[61]

PEN has recommended increasing the authorities of the NEHI working group to empower it to develop and implement the top-down research plan, increasing EHS funding, and appointing a full-time director to support the NEHI working group.

Responding to the PEN recommendation, E. Clayton Teague, director of the NNCO, testified before Congress that there was a consensus among NNI agencies that a centralized office with budgetary authority to oversee the NNI's EHS research program would have significant detrimental effects. According to Dr. Teague,

> No one agency or centralized organization would have the breadth of scientific expertise and knowledge of regulatory authorities and needs currently represented by the 20 agencies participating in the NEHI working group.
>
> Creation of a new central authority would undermine the existing successful interagency coordination.
>
> Moving the management of all nanotechnology EHS research into a single office would likely decouple such research from related efforts within NNI agencies and from the knowledge base in the agencies that is currently networked into the NNI's EHS research effort.
>
> Creating a separate office would, on the one hand, give mission agencies a disincentive for doing nanotechnology-related EHS research. They would reasonably

assume that another agency is responsible, and they therefore could redirect their limited resources to address other priorities. A likely result could be that the level of research would actually decrease. Conversely, creating a separate office could lead to duplicative work being funded, thereby wasting tax dollars and not optimizing progress.[62]

A Cooperative Approach to Addressing EHS Concerns

Some organizations have taken a cooperative approach to promote EHS research. For example, the Environmental Defense Fund, an environmental advocacy group, partnered with the American Chemistry Council, a trade group, to issue a Joint Statement of Principles in June 2005 that recognizes the "significant societal and sustainable development benefits" expected from nanotechnology, while calling for a multi-stakeholder dialogue to achieve the timely development of nanomaterials "in a way that minimizes potential risks to human health and the environment." The statement also called for increased federal investments in EHS research and development of an international effort to standardize testing protocols, hazard and exposure assessment approaches, and nomenclature and terminology ... to maximize resources and minimize inconsistent regulation of nanomaterials.[a]

There is general agreement among stakeholders that these activities can contribute to creating an environment where research results can be reliably shared and compared, to protecting human health and safety, and to creating a common language about nanotechnology that increases clarity in the sharing of ideas and information. However international standardization efforts are often time- and resource-consuming, and can divert resources from more pressing needs. In addition, such efforts can be used by nations and other organizations for competitive advantage (e.g., by securing the adoption of a favorable standard, slowing others' progress).

In June 2007, the Environmental Defense Fund and DuPont issued a Nano Risk Framework "to assist with the responsible development and use of nanotechnology and to help inform global dialogue on its potential risks."[b] The framework is a six-step process to identify, address, and manage potential risks: (1) describe the material and the intended application; (2) profile the material's lifecycle in the application; (3) evaluate associated risks; (4) assess risk management options; (5) decide on and document actions; and, (6) regularly review new information and adapt actions accordingly.[c]

[a] Environmental Defense and American Chemistry Council Nanotechnology Panel: Joint Statement of Principles, Comments on EPA's Notice of a Public Meeting on Nanoscale Materials, June 23, 2005.

[b] "DuPont and Environmental Defense Launch Comprehensive Tool for Evaluating and Addressing Potential Risks of Nanoscale Materials," press release, E. I. du Pont de Nemours and Company, June 21, 2007. http://vocuspr.vocus.com/VocusPR30/Newsroom/Query. aspx?SiteName =Dupont New&Entity=PRAsset&SF_PRAsset_PRAssetID_EQ=106677& XSL=PressRelease &Cache=False

[c] Nanorisk Framework, Environmental Defense-DuPont Nano Partnership, June 2007.

Andrew Maynard counters that "it should be possible to develop a functional structure that enables agencies to work within a broader plan." According to Maynard, while a centralized office is not necessary, top-down leadership with authority and the ability to ensure resources get to where they are needed is necessary.... [Such] leadership does not take away from agencies' expertise and missions, but rather empowers agencies to do the best they can, while coordinating and partnering as effectively as possible with each other.[63]

Project on Emerging Nanotechnologies Recommendations

The Project on Emerging Nanotechnologies (PEN), a joint venture of the congressionally-chartered Woodrow Wilson Center for International Scholars and the Pew Charitable Trusts, has produced inventories of both nanotechnology-based products and government-funded EHS research. PEN has asserted the need for more EHS research, more aggressive oversight, and a more centralized federal government approach to funding EHS research.

In addition, PEN contends that the increasing complexity of systems incorporating nanoparticles with multiple functions will make the behaviors more complex and difficult to predict. To minimize the likelihood of a nanotechnology accident, PEN made the following recommendations:

- Creating a Nano Safety Reporting System where people working with nanotechnology can anonymously report safety issues and concerns. PEN states that the information gleaned from this system could be used to inform the design of educational materials, better structure technical assistance programs, and provide an early indicator of emerging safety issues.
- Creating technologies that provide an early-warning system to allow for risk to be assessed early in research efforts. Such a technology might enable low-cost, fast-screening for novel properties that would allow for risk assessment integrated and concurrent with the R&D process.
- Pushing information out to small businesses, start-ups, and laboratories that, due to their size and resources, are unlikely to be able to devote significant resources to EHS issues. PEN states that existing assistance programs could be used to deliver this information, as well as the development of peer-to-peer mentoring programs within industrial supply chains.
- Application of lessons learned in other technology areas to make nanotechnology more inherently safe, using strategies such as multiple levels of protection, learning from failures, not oversimplifying the complex, awareness of operations, and building in resilience to prevent cascading of errors.

Federal Regulation

Some have raised concerns about whether current laws, regulations, and authorities are adequate to protect human health, safety, and the environment from potential adverse implications of nanotechnology. Several factors may affect the ability of the regulatory system to keep pace with advances in technology, both broadly and specifically with respect to nanotechnology.

Broadly, market forces have increased the pace of global innovation, challenging institutions' ability to identify and cope with the societal implications of rapid change. Speed-to-market has become a driving factor in competition for many industries as a result of the entry of new and nimble competitors in the global marketplace, increased public and private investments in R&D, global models of innovation, increased flows of scientific and technical knowledge, and greater numbers of scientists and engineers around the world. In addition, growing global markets enable companies to recoup their investments faster and enable earlier investments in subsequent generations of technology, further accelerating the pace of innovation. The increased pace, scope, and complexity of technological innovation may pose challenges to the existing regulatory system. While these factors may affect a broad range of technologies, nanotechnology may be especially affected due to the rapid growth in public and private R&D investments in the field since the year 2000 and the potential for nanomaterials to be used in a wide array of products.

Unique Properties Emerge at the Nanoscale

Scientists have discovered that elements and materials with the same chemistry can exhibit fundamentally different properties at the nanoscale. For example, platinum, which exhibits no magnetism in its bulk form, shows significant magnetic properties in nanoscale clusters of 13 atoms. The optical properties of gold also can change with particle size. At 10 nanometers, gold particles absorb green light and appear red, not gold.

Not only can nanoscale particles differ in properties from bulk material with the same chemical composition, they may also differ from other nanoscale materials with the same chemical composition. For example, the melting point of an element—which was believed to be constant regardless of the element's particle size—can change with particle size. Nanotechnology research has demonstrated that the melting temperature of gold decreases when the particle's radius drops below 10 nanometers (from a melting temperature of approximately 1,000oC at 10 nanometers to approximately 500oC at 2 nanometers).

Source: Roduner, Emil. "Nanoscopic Materials: Size-Dependent Phenomena," University of Stuttgart, Germany, August 2006.

Nanotechnology also may pose unique challenges to the regulatory system. For example, historically, regulatory agencies have defined a chemical by its chemical composition, usually without regard to its particle size. In contrast, the essence of nanotechnology is that a material may exhibit different properties at the nanoscale than it does at a bulk, molecular, or atomic scale. (See text box, "Unique Properties Emerge at the Nanoscale.") Accordingly, questions are being raised by representatives of the scientific, advocacy, and regulatory communities about how an EHS research portfolio might be structured when particle size may affect a material's properties, whether it may be necessary to incorporate particle size into regulatory regimes, and how this might be accomplished given the vast spectrum of particle sizes that might affect the characteristics of a particular material.

Some experts argue that EHS concerns about nanotechnology products can be handled under existing laws and regulations, while others see legal obstacles to adequate EHS regulation. In both of its assessments of the NNI, PCAST concluded that existing regulatory authorities were adequate for the current activities; that appropriate regulatory mechanisms should be used to address instances of harmful human or environmental effects of nanotechnology; and that new regulatory policies related to nanotechnology should be rational, science-based, and consistent across the federal government. Similarly, Sean Murdock, then-executive director of the NanoBusiness Alliance, asserted that

> The apparatus for effective nanotechnology regulation is largely in place through various statutes and agencies, but it lacks data and resources. To enable these agencies and for the nanotech regulation effort to succeed we must increase the level of funding available to them for nanotech environmental, health and safety research; coordinate efforts between agencies; establish metrics and standards that can be used to characterize nanomaterials; conduct ongoing research; and more.[64]

Others believe that new laws and regulations, or modifications to existing ones, may be required. J. Clarence Davies, senior advisor to the Project on Emerging Nanotechnologies and former EPA Assistant Administrator for Policy, Planning, and Evaluation argued that

> Nanotechnology is difficult to address using existing regulations. There are a number of existing laws—notably the Toxic Substances Control Act; the Occupational Safety and Health Act; the Food, Drug and Cosmetic Act; and the major environmental laws (Clean Air Act, Clean Water Act, and Resource Conservation and Recovery Act)—that provide some legal basis for reviewing and regulating [nanotechnology] materials. However, all of these laws either suffer from major shortcomings of legal authority, or from a gross lack of resources, or both. They provide a very weak basis for identifying and protecting the public from potential risk, especially as nanotechnologies become more complex in structure and function and the applications become more diverse.
>
> A new law may be required to manage potential risks of nanotechnology. The law would require manufacturers to submit a sustainability plan which would show that the product will not present an unacceptable risk.[65]

In a 2008 PEN report, *Oversight of Next Generation Nanotechnology*, Davies asserted that nanotechnology, along with other advanced technologies, have characteristics that challenge conventional methods of risk assessment, standard setting, and oversight implementation, severely hampering the effectiveness of the existing regulatory structure.

> Since 1980, the capability of the federal agencies responsible for environmental health and safety has steadily eroded. The agencies cannot perform their basic functions now, and they are completely unable to cope with the new challenges they face in the 21st century.[66]

As an alternative, Davies put forward a concept for a Department of Environmental and Consumer Protection, "a scientific agency with a strong oversight component, in contrast to the current regulatory agencies, which are primarily oversight bodies." The agency would incorporate six existing regulatory and science agencies and establish new units for risk assessment, forecasting, technology assessment, health monitoring, and collection of environmental statistics.[67]

Davies also stated that new mechanisms and institutional capabilities—including research programs, tax breaks, acquisition programs, and regulatory incentives—are needed to encourage beneficial applications of nanotechnology.

In developing the regulatory structure, some in the business and financial communities argue that stability and predictability are key characteristics for attracting investment and spurring commercial applications. According to Matthew Nordan, then-vice president of Lux Research, the

> ambiguity surrounding environmental, health, and safety regulation of nanoparticles is hampering commercialization. Firms do not want to play a game whose rules may change at any time.... That doesn't mean they want more regulations or more onerous regulations. They're just looking for a roadmap on how federal agencies such as the EPA or OSHA [Occupational Safety and Health Administration] plan to approach nanoparticles.[68]

Some tension exists between the goals of promoting the development of nanotechnology, ensuring the global competitive position of the United States, addressing potential EHS implications of nanotechnology, and coping with the unique challenges nanotechnology poses to the current regulatory regime. To prevent health and safety concerns from becoming an impediment to innovation, some suggest that health and safety research and regulation must be done near-concurrently with product development, keeping pace with the speed of innovation. Alternatively, others argue that the potential health, safety, and environmental implications are either unknown or of such significance that EHS research and regulation must precede nanotechnology development and commercialization. "By the time monitoring catches up to commerce the damage will already have been done," asserted Ian Illuminato, health and environment campaigner for Friends of the Earth.[69] AFL-CIO industrial hygienist Bill Kojola warned that

> Even though potential health hazards stemming from exposure have been clearly identified, there are no mandatory workplace measures that require exposures to be assessed, workers to be trained, or control measures to be implemented. [Nanotechnology] should not be rushed to market until these failings are corrected and workers assured of their safety.[70]

The National Research Council assessment of the NNI acknowledged the need for additional reproducible, well-characterized EHS data to inform risk-based guidelines and best practices and warned that until such information is available precautionary measures should be taken to protect the health and safety of workers, the public, and the environment.[71]

In its 2008 assessment of the NNI, PCAST asserted that risk research must not be considered in isolation, but rather in the context of the overall risks and benefits of a particular material or technology. This perspective is shared by many industry advocates who argue that regulatory decisions must balance the potential risks associated with a nanotechnology product against the benefits it delivers and the risk it displaces. Further, they maintain that nanotechnology products should not be held to a higher standard than non-nanotechnology products. PCAST also noted that manufacturers and sellers of nanotechnology products had responsibilities for ensuring workplace and product safety, and asserted that the NNI has a vital role in supporting federal regulatory agencies by providing them with EHS research results.

International Engagement

International engagement on EHS issues is believed by many to be important to the responsible development and successful commercialization of nanotechnology. NNI officials assert that the United States has played a central role in convening international efforts to address EHS concerns. In its 2008 assessment, PCAST encouraged the NNI to coordinate its efforts with other nations to avoid duplication and to leverage investments, characterizing such work as "noncompetitive."[72] In its 2010 assessment, PCAST acknowledged the wide range of international engagement by the NNI and its member agencies and recommended that these efforts be "continued and expanded."[73]

Federal agencies have engaged internationally (e.g., with agencies of other nations, international organizations, standards organizations) across a wide range of nanotechnology-related areas, including standards, nomenclature, and EHS research. The NSET established the Global Issues in Nanotechnology (GIN) working group in 2005 to monitor foreign nanotechnology programs, promote U.S. commercial and trade interests in nanotechnology, and broaden international collaboration on nanotechnology R&D, including research on safeguarding the environment and human health.

Advocates for international engagement assert a variety of potential benefits. For example, transparency and/or harmonization of standards and regulations may contribute to assurance of global supply chains and market confidence in nanotechnology products. Increased globalization of production and markets means that companies and consumers around the world are increasingly part of a common network. Manufacturers of final products generally rely on inputs from multiple suppliers in their global supply chains. The reliability of a final product often depends on the reliability of inputs, such as materials or components. Transparent and common standards and regulations may help to ensure the integrity of supply chains and final products. While this is an issue for a variety of non-nanotechnology products (e.g., the recent discovery of lead-tainted toys and other products imported from China), nanotechnology may present a unique challenge in that at least some nanoscale particles can be incorporated into materials and products in ways that cannot be easily detected or detected at all. Thus, producers and the consumers they serve must rely, in large measure, on standards and regulatory systems to ensure that nanoscale materials are properly produced and represented throughout the supply chain. In the absence of such standards and regulatory systems, producers may not be able to rely on inputs or may incur additional costs for testing and verification; substandard inputs may be incorporated in final products making them underperform or unsafe, and possibly resulting in loss of market confidence and/or potential litigation; or nanotechnology materials may be incorporated without disclosure.

Internationally agreed upon standards could also contribute to greater comparability of research results, improving understanding of EHS-related aspects of nanotechnology, and promoting regulations that help protect human health and the environment. Common standards and nomenclature also may contribute to more effective global R&D collaboration, accelerating the realization of nanotechnology's economic and societal potential.

Global engagement may help to establish a common environment for the development and production of nanotechnology products and to promote access to global markets. In the absence of such an environment, some nations may seek to attract investments in their markets by adopting lower environmental, health, and safety standards and regulations.

Finally, while much remains unknown about the transport and fate of nanoscale materials released into the environment, it is possible that countries and populations other than those where research and production activities take place may be affected. Efforts to promote the adoption of best practices in nanotechnology research, production, use, disposal, and recycling may protect human health and the environment worldwide.

International engagement on EHS research may pose problems, including the time, cost, difficulty, and alleged ineffectiveness of such collaborations. For example, while some advocates assert the need for swift action in advancing EHS research, international engagements often entail slow processes. Also, given the strong U.S. position in nanotechnology, broadly, and in nanotechnology EHS research, specifically, some may argue that other countries have little to contribute, that such efforts tax limited federal EHS financial and human resources, and that such diffusion of resources may slow overall EHS progress. Others might assert that international engagement efforts focused explicitly on nano technology are unnecessary given the wide variety of existing mechanisms and pathways for sharing academic research and environmental, health, and safety information across national borders.

Some may oppose international engagement efforts because they lack faith in the goodwill of participating parties due to the potentially strong national interests at stake (e.g., military applications, economic growth, job creation). In 2003, then-Under Secretary of Commerce for Technology Phillip J. Bond questioned whether global calls for a slowdown in nanotechnology R&D to address EHS concerns were intended to allow other nations to close the nanotechnology leadership gap with the United States:

> I wonder very often if there are really calls for a slow-down so that other governments and countries might catch up.[74]

Others assert that the research required to understand and address EHS implications may be closely linked to applications-related R&D to create nanotechnology materials, products, or processes. In such cases, companies and countries may be reluctant to reveal EHS concerns and efforts, to cooperate in EHS research, or to share results as such actions may reveal competitive strategies, provide information others might use to compete against them (e.g., insights into promising materials or manufacturing processes), or result in unwanted scrutiny by regulators.

CONCLUSION

Advocates and critics agree that potential environmental, health, and safety implications of nanotechnology must be addressed if the full economic and societal benefits of nanotechnology are to be achieved. There is also general agreement that the current body of knowledge of how nanoscale materials might affect humans and the environment is insufficient to assess, address, and manage the potential risks. While there is agreement on the need for more EHS research, there are differing views on the level of funding required, how it should be managed, and related issues.

In the 111th Congress a variety of legislation was considered seeking to address, in some manner, EHS-related issues, including: H.R. 5116 (111th Congress) (Title I, Subtitle A); H.R. 554 (111th Congress) and S. 1482 (111th Congress), both titled "National Nanotechnology Initiative Amendments Act of 2009," which would reauthorize and amend the 21st Century Nanotechnology Research and Development Act; S. 2942 (111th Congress), the Nanotechnology Safety Act of 2010; H.R. 820 (111th Congress), the Nanotechnology

Advancement and New Opportunities Act; and the appropriations bills that fund the NNI agencies' nanotechnology EHS research.

None of these bills were enacted. The 112th Congress may again seek to address nanotechnology EHS implications issues, including:

- Is there a need for a national EHS research strategy to identify and address knowledge gaps? If so, which institutions should be a part of such a strategy? Which institution(s) should develop such a strategy?
- Should the federal approach to EHS research be bottom-up, driven by individual agency decisions and coordinated by the NNCO? Should it be top-down with a central controlling authority? Or should the federal government take a hybrid approach, using a central office with its own funding to address research needs not addressed by other agencies?
- How much should the federal government appropriate for EHS research? Should the amount of EHS funding be proportionate to the overall NNI budget? How should the research be prioritized? How can the federal EHS research investment be better accounted for? How can EHS research results and best practices be shared more broadly?
- Can voluntary programs effectively provide needed information about industrial nanotechnology production activities? Are existing laws, regulations, guidelines, and regulatory structures adequate? Should agencies be more aggressive in their use of regulatory authority to collect more information from companies about the nanotechnology and nanotechnology-enabled products they manufacture? Is there sufficient coordination among federal regulatory agencies?
- How can efforts to develop common nomenclature and standards be improved? What types of international engagement on nanotechnology research and regulatory issues could best foster responsible development of nanotechnology and ensure confidence in supply chains?

Congress' approach to each of these issues may have a substantial effect on U.S. leadership in nanotechnology R&D and commercialization, the realization of the potential societal benefits of nanotechnology, public health and safety, the environment, and the public policy decisions and investments made by other nations.

NANOTECHNOLOGY EHS-RELATED LEGISLATION IN THE 111TH CONGRESS

Five bills introduced in the 111th Congress contained provisions that sought to address nanotechnology EHS concerns. The following section summarizes selected EHS-related provisions of these bills.

Title I, Subtitle A, H.R. 5116 (111th Congress)—National Nanotechnology Initiative Amendments Act of 2010

The provisions of Title I, Subtitle A of H.R. 5116 (111th Congress), the National Nanotechnology Initiative Amendments Act of 2010, are nearly identical to H.R. 554 (see "H.R. 554—National Nanotechnology Initiative Amendments Act of 2009" below). H.R. 5116 changes the name of the act from the "National Nanotechnology Initiative Amendments Act of 2009," to "National Nanotechnology Initiative Amendments Act of 2010," and removes the term "interdisciplinary" from a provision establishing "green nanotechnology" research centers.

H.R. 554 (111th Congress)—National Nanotechnology Initiative Amendments Act of 2009

H.R. 554 (111th Congress), the National Nanotechnology Initiative Amendments Act of 2009, was introduced on January 15, 2009, and referred to the House Committee on Science and Technology. On February 11, 2009, the bill was brought to the floor on a motion to suspend the rules and passed by voice vote. The bill was received in the Senate and referred to the Committee on Commerce, Science, and Transportation. This act would revise the 21st Century Nanotechnology Research and Development Act in a variety of ways, several of which specifically address nanotechnology EHS concerns. The legislation:

- directs the National Nanotechnology Coordination Office to develop and maintain a public database of NNI EHS projects, including the agency funding source and funding history;
- requires the National Nanotechnology Advisory Panel (NNAP) to be established as a "distinct entity" (the NNAP's functions are currently performed by the President's Council of Advisors on Science and Technology), and requires the establishment of a subpanel to assess whether societal, ethical, legal, environmental, and workforce concerns are adequately addressed by the NNI;
- directs that the National Research Council, as part of its triennial review of the NNI, evaluate the adequacy of the NNI's efforts to address ethical, legal, environmental, human health, and other appropriate societal concerns;
- requires the designation of an associate director of the White House Office of Science and Technology Policy to serve as Coordinator for Societal Dimensions of Nanotechnology with responsibility for developing an annual research plan for federal nanotechnology EHS activities, monitoring and encouraging agency EHS efforts, and for encouraging agencies to engage in public-private partnerships to support EHS research;
- requires certain interdisciplinary research centers supported under the NNI to include EHS research to develop methods for developing environmentally benign nanoscale products and processes, to foster the transfer of research results to industry, and to provide interdisciplinary study programs to educate scientists and engineers in these methods;

- directs NNI agencies to support the activities of standards setting bodies involved in the development of standards for nanotechnology, including authorizing agency reimbursement of travel costs of scientists and engineers participating in these activities; and
- requires activities supported under the NNI's Education and Societal Dimensions program component area to include environmental, health, and safety education in its informal, pre-college, and undergraduate nanotechnology education efforts.

S. 1482 (111th Congress)—National Nanotechnology Amendments Act of 2009

S. 1482 (111th Congress), the National Nanotechnology Amendments Act of 2009, was introduced on July 21, 2009, and referred to the Senate Commerce, Science, and Transportation Committee. The purpose of the bill is to reauthorize the 21st Century Nanotechnology Research and Development Act and to expand the scope of the National Nanotechnology Program (NNP).

Among its provisions, the bill:

- requires the NNP to solicit and draw upon the perspectives of the industrial community to promote the rapid commercial development of nanoscale-enabled devices, systems, and technologies and to coordinate research in determining the key physical and chemical characteristics of nanoparticles and nanomaterials that may pose environmental, health, and safety risks;
- requires the NNCO and other appropriate agencies and councils to issue guidance to agencies that describes a strategy for transitioning research into commercial products and technologies and how the program will coordinate or conduct research on the environmental, health, and safety issues related to nanotechnology;
- requires each participating agency to provide funds to support the work of the NNCO. Authorizes appropriations to: (1) NIST for the development of nanotechnology standards; and (2) NSF, for use by the NNCO, to develop and maintain a public information database of NNP projects in EHS; education; public outreach; ethical, legal, and other societal issues; and of nanotechnology facilities accessible for use by individuals from academia and industry;
- makes the National Nanotechnology Advisory Panel (NNAP) a distinct entity, and requires the NNAP to establish a subpanel to enable it to assess whether societal, ethical, legal, environmental, and workforce concerns are adequately addressed by the NNP;
- requires the designation of a "coordinator for societal dimensions of nano technology," within OSTP, to convene a panel to develop a research plan, and requires the coordinator to enter into an arrangement with the National Science Board to create a report that identifies the broad goals and needs of EHS researchers;
- directs the NSTC to establish an interagency Education Working Group to coordinate, prioritize, and plan formal and informal educational activities supported under the NNP, including activities to help participants understand the EHS implications of nanotechnology; and

- requires the NNP to support nanotechnology R&D in areas of national importance (e.g., economic competitiveness, energy production, water purification, agriculture, and health care; in environmental, health, and safety research on the risks of nanoparticles) and in ethical, legal, and societal issues related to nanotechnology.

S. 2942 (111th Congress)—Nanotechnology Safety Act of 2010

S. 2942 (111th Congress), the Nanotechnology Safety Act of 2010, was introduced on January 21, 2010, and referred to the Senate Committee on Health, Education, Labor, and Pensions. The bill would require the Secretary of Health and Human Services to establish within 180 days a program for the scientific investigation of nanoscale materials included or intended for inclusion in FDA-regulated products, to address the potential toxicology of such materials, the effects of such materials on biological systems, and interaction of such materials with biological systems. The bill would authorize $25 million per year for fiscal years 2011 to 2015.

H.R. 820 (111th Congress)—Nanotechnology Advancement and New Opportunities Act

H.R. 820 (111th Congress), the Nanotechnology Advancement and New Opportunities Act, was introduced on February 3, 2009, and referred to the House Science and Technology Committee; the House Ways and Means Committee; the House Energy and Commerce Committee; and the House Homeland Security Committee. Among its provisions, the bill would require the NNCO to produce an annual research strategy that establishes priorities for the development and responsible stewardship of nanotechnology, as well as providing recommendations regarding the funding required to implement the strategy.

Appendix. Overview of Selected Federal Agencies' Roles in the Regulation of Nanotechnology

Several federal regulatory agencies have begun to grapple with the EHS issues raised by nanotechnology in their spheres of responsibility. Some critics argue that there is a potential conflict of interest among some regulatory agencies that are, on the one hand, conducting and promoting nanotechnology research and that are, on the other hand, responsible for regulating nanotechnology applications. The following section provides an overview of selected federal agencies' roles in the regulation of nanotechnology.

Environmental Protection Agency

The Environmental Protection Agency (EPA) has both a research function and a regulatory function. The agency has asserted a need for more information to assess the potential EHS impacts of most engineered nanoscale materials. According to EPA, this information is needed

> ... to establish a sound scientific basis for assessing and managing unreasonable risks that may result from the introduction of nanoscale materials into the environment.[75]

In this regard, EPA is supporting research on the toxicology, fate, transport, transformation, bioavailability, and exposure of humans and other species to nanomaterials to obtain information for use in risk assessment, a central aspect of EPA's mission.[76]

EPA plays a central role in coordinating the federal governments research efforts to address nanotechnology EHS issues, serving as co-chair of the NSET Nanotechnology Environmental Health Implications (NEHI) working group. The National Institute for Occupational Safety and Health (NIOSH), a research institute within the Department of Health and Human Services, is EPA's co-chair of the NEHI working group.

EPA also works with international organizations engaged in nanotechnology-related regulatory issues, such as the International Organization for Standardization and the Organization for Economic Cooperation and Development.

With respect to its regulatory function, multiple statutes govern EPA's authority to regulate nanotechnology materials and devices, including the Clean Air Act (CAA, 42 U.S.C. 7401 et seq); Clean Water Act (CWA, codified generally as 33 U.S.C. §§1251-1387); Federal Insecticide, Fungicide, and Rodenticide Act (FIFRA, 7 U.S.C.136-136y); and Toxic Substances Control Act (15 U.S.C. 2601 et seq.).[77]

Important issues have been raised about the application of EPA's authorities to regulate nanotechnology. Several issues revolve around TSCA, which authorizes regulation of chemical commerce.[78] Under the provisions of TSCA, producers of a "new" material must provide EPA with a premanufacture notification (PMN). EPA then has 90 days to approve manufacture, to require information from manufacturers, or to restrict chemical use. Other TSCA provisions permit EPA regulation of existing chemicals already in commerce, but these rely on EPA fact-finding and rulemaking before EPA can require testing or restrict uses. Several NGOs have urged EPA to consider all nanoscale materials "new" regardless of whether the material is on the EPA inventory list in its bulk form.[79] However, some nanotechnology materials have the same chemical composition as materials that are already in commerce, raising the question of whether the nanotechnology materials are "new" and thus subject to PMN requirements.

When nanomaterials are intended to control pests, including microbes, FIFRA may offer EPA more authority to regulate nanotechnology than TSCA, according to Lynn Bergeson, chair of the American Bar Association's Section on Environment, Energy, and Resources:

> Under TSCA, once a substance is on the approved inventory list, any use is legitimate, but FIFRA is use-specific. The EPA always has the authority to assess the risk of pesticides, regardless of the use.[80]

Applicability of FIFRA to nanotechnology products was one aspect of a November 2006 EPA ruling that a device that "incorporates a substance intended to prevent, destroy or mitigate pests" is considered a pesticide and is required to be registered under FIFRA. While the ruling is not unique to nanomaterials, it came in the context of advertising claims for a washing machine containing nanoscale silver ions that kill microbes. EPA's ruling made this appliance the first nanotechnology product to be regulated under FIFRA. However, claims for the pesticidal effectiveness of the washing machine have been removed from advertisements, possibly limiting EPA's ability to regulate the device as a pesticide under FIFRA.

In a May 2010 review of EPA's role in regulating nanotechnology, the U.S. Government Accountability Office concluded that EPA was missing opportunities to collect additional information under TSCA, FIFRA, the Clean Water Act and other environmental statutes.[81]

Food and Drug Administration

A variety of current and future products that incorporate nanotechnology fall, or may fall, under the regulatory auspices of the FDA, including cosmetics, medical devices, foods, drugs, biological products, and combination products.[82] FDA anticipates that many of the nanotechnology products that the agency is likely to regulate will be combination products, such as drug-device, drug-biological, or device-biological products. According to FDA, it regulates products based on their statutory classification rather than the technology they employ, thus the agency may not provide regulatory consideration to a nanotechnology product until well after its initial development.[83] Also, some critics maintain that FDA's limited regulatory authority over certain categories of products may limit its authority to regulate nanotechnology products.

With respect to the need for unique tests or requirements for regulating nanotechnology products, FDA states that its existing requirements may be adequate for most nanotechnology products it expects to regulate. FDA asserts that nanotechnology products are in the same size-range as the cells and molecules its reviewers and scientists deal with every day. The agency says that every degradable medical device and injectable pharmaceutical generates particulates that pass through the nanoscale size range during the processes of their absorption and elimination by the body. FDA says that it has no knowledge of reports of adverse reactions related to the "nano" size of resorbable drug or medical device products. New tests or other requirements may be needed, according to FDA, if new risks are identified arising from new materials or manufacturing techniques. Others, in particular consumer groups, counter that FDA's resources are insufficient to adequately address the safety of emerging technologies in general, and that the agency's regulatory approach, particularly for cosmetics, dietary supplements, and other products for which pre-market review is not required, would not detect any problems until such products had been in use.[84]

FDA does not provide grants for nanotechnology research but does conduct research in several of its centers to understand the characteristics of nanomaterials and nanotechnology processes. FDA is collaborating with NIEHS on studies, as part of the interagency National Toxicology Program.

FDA says that there currently is no international regulation of nanoproducts or the underlying nanotechnology. FDA participates in multinational organizations where cooperative work on nanotechnology has been proposed, including the Organization for Economic Cooperation and Development, ASTM International, and the International Organization for Standardization.

National Institute of Environmental Health Sciences/National Toxicology Program

While not a regulatory agency, NIEHS, a part of the National Institutes of Health, is conducting nanotechnology EHS research that will support the missions of regulatory agencies. In particular, NIEHS serves as home to the interagency National Toxicology Program (NTP). The NTP's mission is to coordinate toxicological testing programs, develop and validate improved testing methods, develop approaches and generate data to strengthen scientific knowledge about potentially hazardous substances, and communicate with stakeholders.[85] In 2006, the NTP established the Nanotechnology Safety Initiative (NSI), a broad-based research program to address potential human health hazards associated with the manufacture and use of nanoscale materials. The goal of this research program is to evaluate the toxicological properties of major nanoscale materials that represent a cross-section of composition, size, surface coatings, and physical and chemical properties, and to use these as model systems to investigate fundamental questions concerning whether nanoscale materials can interact with biological systems and how they might do so.[86]

According to NTP, the NSI is focused on three areas of research with respect to specific types or groups of nanoscale materials:

- non-medical, commercially relevant and available nanoscale materials to which humans are intentionally being exposed, such as cosmetics and sunscreens;
- nanoscale materials representing specific classes (e.g., fullerenes and metal oxides) so that information can be extrapolated to other members of those classes; and
- subsets of nanomaterials to test specific hypotheses about a key characteristic (such as size, composition, shape, or surface chemistry) that might be related to biological activity.[87]

NSI research activities are focused on metal oxides, fluorescent crystalline semi conductors (also known as quantum dots), fullerenes, carbon nanotubes, nanoscale silver, and nanoscale gold.

Occupational Safety and Health Administration/National Institute for Occupational Safety and Health

The mission of the Occupational Safety and Health Administration (OSHA), an agency of the Department of Labor, is to ensure the safety and health of America's workers by setting and enforcing standards; providing training, outreach, and education; establishing partnerships; and encouraging continual improvement in workplace safety and health. OSHA has not yet taken any regulatory actions with respect to nanotechnology.

The National Institute for Occupational Safety and Health (NIOSH), a part of the Centers for Disease Control, is the lead federal agency conducting research and providing guidance on the occupational safety and health implications and applications of nanotechnology. NIOSH co-chairs the NSET's NEHI working group together with EPA. NIOSH is not a regulatory agency, but its work directly supports OSHA and other regulatory agencies. NIOSH and

OSHA are considering new risk management approaches that seek to maximize flexibility for innovation while ensuring the health and safety of workers.[88]

NIOSH states that its nanotechnology efforts are building on its experience in defining the characteristics, properties, and effects of ultrafine particles—such as welding fumes and diesel particulates—as well as its experience in conducting advanced health effects laboratory studies and in fostering industrial hygiene policies and practices.

NIOSH has developed interim guidelines for working with nanomaterials. The agency asserts that these guidelines are consistent with the best scientific knowledge of nanoparticle toxicity and control. NIOSH also maintains a Nanoparticle Information Library with information on the health and associated properties of nanomaterials as an online resource for occupational health professionals, industrial users, worker groups, and researchers.[89]

Consumer Product Safety Commission

The Consumer Product Safety Commission (CPSC) is charged with protecting the public from unreasonable risks of serious injury or death from certain types of consumer products.[90] CPSC has asserted that potential safety and health risks of nanomaterials can be assessed under existing CPSC statutes, regulations and guidelines. Since the Consumer Product Safety Act (15 U.S.C. 2051 et seq.) and the Federal Hazardous Substances Act (15 U.S.C. 1261 et seq.) do not require pre-market registration or approval of products, CPSC does not evaluate a product's risk to the public until it has been distributed in commerce.

In August 2005, CPSC commissioners approved a nanotechnology statement which notes that nanotechnology presents challenges that "may require unique exposure and risk assessment strategies." The CPSC statement identified regulatory challenges, including identification of the specific nanomaterial in a product; the need to characterize the materials to which a consumer is exposed during product use, including an assessment of the size distribution of the materials released; and the application of toxicological data of appropriate particle sizes to assess health risks. The CPSC takes the position that it is unable to make any general statements about potential consumer exposure to nanomaterials or the health effects that may result from exposure to nanomaterials during consumer use and disposal due to the wide variation in potential health effects and the dearth of exposure and toxicity data for specific nanomaterials.[91]

End Notes

[1] Congress defined nanotechnology in the 21st Century Nanotechnology Research and Development Act (P.L. 108-153) as, "the science and technology that will enable one to understand, measure, manipulate, and manufacture at the atomic, molecular, and supramolecular levels, aimed at creating materials, devices, and systems with fundamentally new molecular organization, properties, and functions." ASTM International, one of the largest voluntary standards development organizations, has defined nanotechnology as, "a wide range of technologies that measure, manipulate, or incorporate materials and/or features with at least one dimension between approximately 1 and 100 nanometers. Such applications exploit those properties, distinct from bulk or molecular systems, of nanoscale components." One nanometer is about the width of 10 hydrogen atoms placed side-by-side, or approximately 1/100,0000 of the thickness of a sheet of paper.

[2] Nanotechnology EHS applications refers to the beneficial use of nanotechnology to improve health, safety, and the environment; EHS implications refers to known and potential adverse effects of nanoscale materials on health, safety, and the environment.

[3] Project on Emerging Nanotechnologies. Figure as of June 2010.

[4] Lux Research, as cited by *Report to the President and Congress on the Third Assessment of the National Nanotechnology Initiative*, President's Council of Advisors on Science and Technology, The White House, March 12, 2010, p. 25.

[5] "Genetically Modified Crops and Foods," Friends of the Earth, January 2003. http://www.foe.co.uk/resource/briefings/gm_crops_food.pdf

[6] "Fact Sheet for Nanotechnology under the Toxic Substances Control Act," Environmental Protection Agency. http://www.epa.gov/oppt/nano/nano-facts.htm

[7] "Cancer Nanotech Plan Gets Nod of Approval," Science, Vol. 305, July 23, 2004. http://www.sciencemag.org/content/vol305/issue5683/s-scope.dtl#305/5683/461c

[8] *A Matter of Size: Triennial Review of the National Nanotechnology Initiative,* National Research Council, 2006. p. 148.

[9] *Nanosilver: A Threat to Soil, Water and Human Health*? Friends of the Earth, March 2007. http://www.foe.org/pdf/FoE_Australia_Nanosilver_report.pdf

[10] "Blood-Brain Barrier Breached by New Therapeutic Strategy," press release, National Institutes of Health, June 2007. http://www3.niaid.nih.gov/news/newsreleases/2007/bloodbrainbarrier.htm

[11] "Nanotechnology Risks: How Buckyballs Hurt Cells," Science Daily, May 27, 2008. http://www.sciencedaily.com/releases/2008/05/080527091910.htm

[12] *National Nanotechnology Initiative: Research and Development Supporting the Next Industrial Revolution, Supplement to the President's FY2004 Budget,* Nanoscale Science, Engineering and Technology Subcommittee, National Science and Technology Council, The White House, October 2003. http://www.nano.gov/ nni04_budget_supplement.pdf

[13] Magrez, A., Kasas, S., Salicio, V., Pasquier, N., Seo, J.W., Celio, M., Catsicas, S., Schwaller, B., and Forro, L. "Cellular Toxicity of Carbon-Based Nanomaterials," *Nano Letters*, 6(6):1121-1125, American Chemical Society, May 2006. http://pubs.acs.org/cgi-bin/abstract.cgi/nalefd/2006/6/i06/abs/nl060162e.htmll

[14] *Nanotechnology: The Future is Coming Sooner Thank You Think,* Joint Economic Committee, U.S. Congress, March 2007. p. 13. http://www.house.gov/jec/publications/110/nanotechnology

[15] The 21st Century Nanotechnology Research and Development Act (P.L. 108-153) requires a triennial assessment of the National Nanotechnology Program (in practice, of the NNI) by the NRC and a biennial assessment by PCAST, serving in its capacity as the National Nanotechnology Advisory Panel (NNAP). The act requires each assessment to include a review of the NNI's EHS activities. Four such assessments have been conducted, one by the NRC (*A Matter of Size: Triennial Review of the National Nanotechnology Initiative*, 2006) and three by PCAST (*The National Nanotechnology Initiative at Five Years: Assessments and Recommendations of the National Nanotechnology Advisory Panel*, May 2005; *The National Nanotechnology Initiative: Second Assessment and Recommendations of the National Nanotechnology Advisory Panel*, April 2008; and *Report to the President and Congress on the Third Assessment of the National Nanotechnology Initiative*, March 2010). In addition, in 2009 the NRC produced a report at the request of the National Nanotechnology Coordination Office entitled, *Review of Federal Strategy for Nanotechnology-related Environmental, Health, and Safety (EHS) Research.*

[16] *A Matter of Size: Triennial Review of the National Nanotechnology Initiative,* National Research Council, 2006. p. 90.

[17] *The National Nanotechnology Initiative: Second Assessment and Recommendations of the National Nanotechnology Advisory Panel,* President's Council of Advisors on Science and Technology, The White House, April 2008. p. 7.

[18] *Report to the President and Congress on the Third Assessment of the National Nanotechnology Initiative,* President's Council of Advisors on Science and Technology, The White House, March 12, 2010, p. 38. http://www.whitehouse.gov/sites/default/files/microsites/ostp/pcast-nano-report.pdf

[19] Bond, Phillip J., Under Secretary for Technology, U.S. Department of Commerce. "Preparing the Path for Nanotechnology: Addressing Legitimate Societal and Ethical Issues," keynote address, Nanoscale Science, Engineering, and Technology Subcommittee Workshop on Societal Implications of Nanoscience and Nanotechnology, December 3, 2003.

[20] *Report to the President and Congress on the Third Assessment of the National Nanotechnology Initiative,* President's Council of Advisors on Science and Technology, The White House, March 12, 2010, p. 38.

[21] *Review of Federal Strategy for Nanotechnology-related Environmental, Health, and Safety (EHS) Research,* National Research Council, Washington, DC, December 2008.

[22] "Survey of U.S. Nanotechnology Executives," Small Times Magazine and the Center for Economic and Civic Opinion at the University of Massachusetts-Lowell, Fall 2006.

[23] Rejeski, David, Director, Project on Emerging Nanotechnologies. "Nanotech Safety 101 or How to Avoid the Next Little Accident," paper, Workshop on Disaster Prevention, Harvard University, April 27, 2006. http://www.nanotechproject.org/file_download/files/nanotechsafety101paper.pdf

[24] NEHI is a working group of the Nanoscale Science, Engineering, and Technology (NSET) Subcommittee of the White House National Science and Technology Council (NSTC). The NSET Subcommittee is the coordinating body for the NNI. For additional information about the structure of the NNI, see CRS Report RL34401, *The National Nanotechnology Initiative: Overview, Reauthorization, and Appropriations Issues.*

[25] According to the NNCO, EHS research funding data included in Tables 1 and 2 of this report are for implications research only. The NNCO also states that the figures reported in Table 1 may understate the NNI's EHS implications research by excluding funding for instrument research, metrology, and standards that support EHS implications research but are reported separately. (Source: Private communication between the NNCO and CRS.)

[26] Regular FY2009 appropriations only; does not include supplemental funding provided under the American Recovery and Reinvestment Act of 2009 (P.L. 111-5).

[27] *The National Nanotechnology Initiative: Second Assessment and Recommendations of the National Nanotechnology Advisory Panel,* President's Council of Advisors on Science and Technology, The White House, April 2008. p. 34.

[28] *Prioritization of Environmental, Health, and Safety Research Needs for Engineered Nanoscale Materials,* Nanoscale Science, Engineering, and Technology Subcommittee, National Science and Technology Council, The White House, August 2007.

[29] *Environmental, Health, and Safety Research Needs for Engineered Nanoscale Materials,* Nanoscale Science, Engineering, and Technology Subcommittee, National Science and Technology Council, The White House, September 2006. Cover letter.

[30] Teague, E. Clayton, director, National Nanotechnology Coordination Office. Testimony before the Subcommittee on Research and Science Education, Committee on Science and Technology, U.S. House of Representatives. Hearing on "Research on Environmental and Safety Impacts of Nanotechnology: Current Status of Planning and Implementation under the National Nanotechnology Initiative." 110th Cong., 1st Sess., October 31, 2007.

[31] *A Matter of Size: Triennial Review of the National Nanotechnology Initiative,* National Research Council, 2006. p. 92.

[32] *The National Nanotechnology Initiative: Second Assessment and Recommendations of the National Nanotechnology Advisory Panel,* President's Council of Advisors on Science and Technology, The White House, April 2008. pp. 7, 27.

[33] *The National Nanotechnology Initiative: Third Assessment and Recommendations of the National Nanotechnology Advisory Panel,* President's Council of Advisors on Science and Technology, The White House, March 12, 2010. p. 45.

[34] Ibid.

[35] Murdock, Sean, executive director, NanoBusiness Alliance. Testimony before the Committee on Science and Technology, U.S. House of Representatives. Hearing on "The National Nanotechnology Initiative Amendments Act of 2008." 110th Cong., 2nd Sess., April 16, 2008.

[36] Denison, Richard A. "A Proposal to Increase Federal Funding of Nanotechnology Risk Research to at least $100 Million Annually," Environmental Defense, April 2005. http://www.edf.org/documents/4442_100 milquestionl.pdf

[37] *The National Nanotechnology Initiative: Second Assessment and Recommendations of the National Nanotechnology Advisory Panel,* President's Council of Advisors on Science and Technology, The White House, April 2008. pp. 7, 27.

[38] Kvamme, Floyd, co-chair, President's Council of Advisors on Science and Technology. Testimony before the Subcommittee on Research and Science Education, Committee on Science and Technology, U.S. House of Representatives. Hearing on "Research on Environmental and Safety Impacts of Nanotechnology: Current Status of Planning and Implementation under the National Nanotechnology Initiative," 110th Cong., 1st Sess., October 31, 2007.

[39] Maynard, Andrew, chief science advisor, Project on Emerging Nanotechnologies, a joint venture of the congressionally-chartered Woodrow Wilson Center for International Scholars and the Pew Charitable Trusts.

Testimony before the Subcommittee on Research and Science Education, Committee on Science and Technology, U.S. House of Representatives. Hearing on "Research on Environmental and Safety Impacts of Nanotechnology: Current Status of Planning and Implementation under the National Nanotechnology Initiative," 110th Cong., 1st Sess., October 31, 2007.

[40] The precautionary principle has been used in other countries on some issues and is the official policy in the European Union. For international agreements a precautionary approach is sometimes embraced. For example, the Biosafety Protocol to the 1992 Convention on Biological Diversity incorporates provisions applying the precautionary principle to the safe handling, transfer, and trade of genetically modified organisms. For further information, see CRS Report RL30594, *Biosafety Protocol for Genetically Modified Organisms: Overview*, by Alejandro E. Segarra and Susan R. Fletcher.

[41] "NGOs urge precautionary principle in use of nanomaterials," EurActiv.com, June 14, 2007, http://www.euractiv.com/en/environment nanomaterials/article-164619; Sass, Jennifer, "Nanotechnology and the Precautionary Principle," presentation, Natural Resources Defense Council, 2006. http://docs. nrdc.org/ health

[42] *Nanomaterials, Sunscreens, and Cosmetics: Small Ingredients, Big Risks*, Friends of the Earth, May 2006. http://www.foe.org/camps/comm/nanotech/nanocosmetics.pdf

[43] "No Small Matter," Communique, ETC Group, May/June 2002. http://www.etcgroup.org/upload/publication/ pdf_file/192 The ETC group is a non-governmental organization focused on the global societal impacts of emerging technologies.

[44] "No Small Matter II: The Case for a Global Moratorium," Occasional Paper Series, ETC Group, April 2003. http://www.etcgroup.org/upload/publication/pdf_file/165

[45] *Environmental, Health, and Safety Research Needs for Engineered Nanoscale Materials,* Nanoscale Science, Engineering, and Technology Subcommittee, National Science and Technology Council, The White House, September 2006. p. vii.

[46] Bond, Phillip J., Under Secretary for Technology, U.S. Department of Commerce. "Nanotechnology: Economic Opportunities, Societal and Ethical Challenges," keynote address, NanoCommerce 2003, December 9, 2003.

[47] *A Matter of Size: Triennial Review of the National Nanotechnology Initiative,* National Research Council, 2006. p. 92.

[48] *The National Nanotechnology Initiative at Five Years: Assessments and Recommendations of the National Nanotechnology Advisory Panel*, President's Council of Advisors on Science and Technology, The White House, May 2005. p. 35.

[49] *The National Nanotechnology Initiative: Second Assessment and Recommendations of the National Nanotechnology Advisory Panel*, President's Council of Advisors on Science and Technology, The White House, April 2008. p. 2.

[50] *Federal Strategy for Environmental, Health, and Safety (EHS) Research Needs for Engineered Nanoscale Materials*, Nanoscale Science, Engineering, and Technology Subcommittee, National Science and Technology Council, The White House, February 2008.

[51] *Review of Federal Strategy for Nanotechnology-related Environmental, Health, and Safety (EHS) Research*, National Research Council, Washington, DC, December 2008.

[52] Ibid.

[53] Rejeski, David, director, Project on Emerging Nanotechnologies. Public comments on the Nanoscale Science, Engineering, and Technology Subcommittee's report, *Prioritization of Environmental, Health, and Safety Research Needs for Engineered Nanoscale Materials: An Interim Document for Public Comment*, September 12, 2007. http://www.nanotechproject.org/process/files/5891/nehi_comments_070912_final.pdf

[54] Hearing on "Research on Environmental and Safety Impacts of Nanotechnology: Current Status of Planning and Implementation under the National Nanotechnology Initiative." 110th Cong., 1st Sess., October 31, 2007.

[55] An electronic copy of this letter, dated February 22, 2007, was provided to the Congressional Research Service (CRS) by the American Chemistry Council.

[56] Incorporated as division F of the Consolidated Appropriations Act, 2008 (P.L. 110-161).

[57] S.Rept. 110-91, p. 54.

[58] *Study in Progress: A Research Strategy for Environmental, Health, and Safety Aspects of Engineered Nanomaterials,* National Academies website. http://dels.nas.edu/Study-In-Progress/Research-Strategy-Environmental-Health/BEST-K08-01-A

[59] Ibid.

60 *The National Nanotechnology Initiative: Third Assessment and Recommendations of the National Nanotechnology Advisory Panel*, President's Council of Advisors on Science and Technology, The White House, March 12, 2010. p. xiii.

61 Maynard, Andrew, chief science advisor, Project on Emerging Nanotechnologies. "Public Meeting on Research Needs and Priorities Related to Environmental, Health, and Safety Aspects of Engineered Nanoscale Materials," comments, January 4, 2007. http://www.nano.gov/html/meetings/ehs/uploads/ 20070103_1505_ Nanotechnology_Maynard_NNCO_Comments.pdf

62 Teague, E. Clayton, director, National Nanotechnology Coordination Office. Testimony before the Subcommittee on Research and Science Education, Committee on Science and Technology, U.S. House of Representatives. Hearing on "Research on Environmental and Safety Impacts of Nanotechnology: Current Status of Planning and Implementation under the National Nanotechnology Initiative." 110th Cong., 1st Sess., October 31, 2007.

63 E-mail communication, November 21, 2007.

64 "Nanotechnology Leaders to Converge in Washington, D.C., This Week for NanoBusiness Alliance Public Policy Tour," article, nanotechwire.com, February 16, 2006. http://nanotechwire.com/news.asp?nid=2929

65 Davies, J. Clarence. *Managing the Effects of Nanotechnology*, Project on Emerging Nanotechnologies, Woodrow Wilson International Center for Scholars, Washington, DC, January 2006. p. 3. http://www.nanotech project.org/ process/assets/files/2708/30jen2_mngeffects.pdf

66 Davies, J. Clarence. *Oversight of Next Generation Nanotechnologies*, Project on Emerging Nanotechnologies, Woodrow Wilson International Center for Scholars, Washington, DC, April 2009. http://www. nanotechproject.org/ process/assets/files/7316/pen-18.pdf

67 Ibid.

68 "U.S. Risks Losing Nano Lead," article, physorg.com, July 6, 2005. http://www.physorg.com/news4963.html

69 "International Coalition Calls for Oversight of Nanotechnology," press release, Friends of the Earth, July 31, 2007. http://action.foe.org/dia/organizationsORG/foe/pressRelease.jsp?press_release_KEY=248

70 Ibid.

71 *A Matter of Size: Triennial Review of the National Nanotechnology Initiative*, National Research Council, 2006. p.11.

72 *The National Nanotechnology Initiative: Second Assessment and Recommendations of the National Nanotechnology Advisory Panel*, President's Council of Advisors on Science and Technology, April 2008. p. 33.

73 *Report to the President and Congress on the Third Assessment of the National Nanotechnology Initiative*, President's Council of Advisors on Science and Technology, The White House, March 12, 2010, p. 42.

74 *Regional, State, and Local Initiatives in Nanotechnology*, Nanoscale Science, Engineering, and Technology Subcommittee, National Science and Technology Council, The White House, 2005. p. 33.

75 "Fact Sheet for Nanotechnology under the Toxic Substances Control Act," Environmental Protection Agency. http://www.epa.gov/oppt/nano/nano-facts.htm

76 "Exploratory Research: Nanotechnology Research Grants Investigating Fate, Transport, Transformation, and Exposure of Engineered Nanomaterials: A Joint Research Solicitation - EPA, NSF, & DOE," Environmental Protection Agency. http://es.epa.gov/ncer/rfa/2007/2007_star_nanotech.html

77 For additional information, see CRS Report RL30798, *Environmental Laws: Summaries of Major Statutes Administered by the Environmental Protection Agency*, coordinated by David M. Bearden.

78 For more information about TSCA and nanotechnology, see CRS Report RL34118, *The Toxic Substances Control Act (TSCA): Implementation and New Challenges*, by Linda-Jo Schierow.

79 "The EPA's Toxic Substances Control Act: What you must know," *Small Times*, September/October 2007.

80 "EPA Regulates Nano Product, Not Nano Industry," *Small Times*, January 2007.

81 Ibid.

82 For additional information, see CRS Report RL34334, *The Food and Drug Administration: Budget and Statutory History, FY1980-FY2007*, coordinated by Judith A. Johnson.

83 "FDA and Nanotechnology Products," Food and Drug Administration. http://www.fda.gov/nanotechnology/ faqs.html

84 See, for example, Michael Taylor, *Regulating the Products of Nanotechnology: Does FDA Have the Tools It Needs?* The Project on Emerging Nanotechnologies, October 2006, at http://www.nanotechproject.org/ news/archive/ is_fda_nanotech-ready. The FDA Science Board, Subcommittee on Science and Technology, designated nanotechnology as one of eight emerging technologies that are most challenging for FDA. See FDA Science Board, Subcommittee on Science and Technology, *FDA Science and Mission at Risk*, November

2007, p. 4, at http://www.fda.gov/ohrms/dockets/ac/07/briefing/2007-4329b_02_01_FDA Report on Science and Technology.pdf. The FDA Science Board is the advisory board to the FDA Commissioner.

[85] "Toxicology in the 21st Century: The Role of the National Toxicology Program," Update, National Toxicology Program, January 2004. http://ntp.niehs.nih.gov/ntp/htdocs/Liaison/2004JanLO_News.pdf

[86] *Toxicology in the 21st Century: The Role of the National Toxicology Program,* Department of Health and Human Services, February 2004. http://ntp.niehs.nih.gov/ntp/main_pages/NTPVision.pdf

[87] National Toxicology Program website, available at http://ntp.niehs.nih.gov/?objectid=30302D16-F1F6-975E-7B315D93D4A1246F

[88] *Environmental, Health, and Safety Research Needs for Engineered Nanoscale Materials,* Nanoscale Science, Engineering, and Technology Subcommittee, National Science and Technology Council, The White House, September 2006.

[89] "Nanotechnology at NIOSH," National Institute for Occupational Safety and Health. http://www.cdc.gov/niosh/topics/nanotech/

[90] CPSC's regulatory authorities are provided by the Consumer Product Safety Act; Federal Hazardous Substances Act of 1960, as amended by the Toy Safety Acts of 1969 and 1984 and the Child Protection Amendments of 1966; Poison Prevention Packaging Act of 1970; Flammable Fabrics Act of 1953; and Refrigerator Safety Act of 1956.

[91] "CPSC Nanomaterial Statement," Consumer Product Safety Commission, August 2005. http://www.c psc.gov/library/cpscnanostatement.pdf

INDEX

C

F

G

H

I

J

K

L

M

N

O

P

Q

R

U

V

W

Y